THE ART OF NEUROSCIE

The Art of Neuroscience in Everything was the first book by the celebrated Neuroscientist Abhijit Naskar. Beginning with this book, he has revealed to the world, the untapped potential of the human brain. Today the work of this self-educated scientist has made it possible for us to meet face to face with our inner selves.

THE ART OF NEUROSCIENCE IN EVERYTHING

ABHIJIT NASKAR

Copyright © 2015 Abhijit Naskar

This is a work of non-fiction

All rights reserved. No part of this publication may be reproduced, distributed, or transmitted in any form or by any means, including photocopying, recording, or other electronic or mechanical methods, without the prior written permission of the author, except in the case of brief quotations embodied in critical reviews and certain other noncommercial uses permitted by copyright law.

An Amazon Publishing Company, 2nd Edition, 2015

Printed in United States of America

ISBN-13: 978-1511693233

Dedicated to all the young minds of the world with a burning desire to pursue science

Table of Contents

Preface. 1

Chapter 1

You Drive Me Crazy Like Hell - The Neurobiology of Lust, Attraction, Love and Attachment..4

Chapter 2

Connectivity of Minds - Think of The Friend and The Friend Appears 26

Chapter 3

Science of Empathy & Learning - The Mirror Neurons 38

Chapter 4

Meaning of Dreams 50

Chapter 5

Studies in Hysteria - Emotional Suppression and Its Dangerous Consequences...............................67

Chapter 6

Earth-Brain Bondage - Psychological & Physiological Changes during Fluctuations in Geophysical Parameters 82

Chapter 7

GOD is A Figment of Your Imagination – Experiencing GOD in The Brain..........................95

Bibliography...110

Preface

Mankind is the most mysterious of all species on our beloved planet earth. Its every behavior, every action and every single phrase it utters, undeniably show the excellence, with which Mother Earth has molded the human species throughout the entire evolutionary period.

Modern man has come a long way, since he first learnt to control and create fire. He has advanced in various fields like science, arts and technology. Even so, it's the privilege of only a neuroscientist to observe and explore the biological or more specifically neurological functions behind all those advancements. My work is to find out what makes the humans so special and unique among all the species on this planet.

Humans are the only species on earth who can contemplate the vastness of the cosmos, the beauty of the sea, peacefulness of the full moon, the craftsmanship of Mother Nature and even contemplate itself contemplating. We can observe the craftsmanship of Mother Nature most significantly in the evolution of human brain. This 3 lbs. lump of jelly is the driving force, the alpha and the omega of all human excellence.

With this tiny piece of biological instrument we are even able to see galaxies that are light-years away from our dearest Milky Way. Alongside such scientific excellence, most of humanity has also accepted the existence of a Supreme Being, which is usually referred to as "GOD", based on faith and historical records of encounters. This gives us an amazing evolutionary trait or behavioral characteristic called "Faith" to explore.

All experiences, behaviors, beliefs, feelings such as faith, love, attraction, lust, hatred, excitement, kindness, empathy, good and evil that make us humans, are the creation of various intricate and inexplicable molecular interaction within the brain. This book is about those interactions that impact over daily human activity and behaviors. In this book, I'll open up to the reader, the beautiful maze of the brain that creates human experiences, feelings and beliefs in the most non-technical way possible. This book will elaborate on the biological or rather neurological foundation of human mind's deepest instincts, emotions and mysteries, and make it coherently understandable even to the layman.

Chapter 1

You Drive Me Crazy Like Hell - The Neurobiology of Lust, Attraction, Love and Attachment

Love is patient and kind. Love is not jealous or boastful or proud or rude. It does not demand its own way. It is not irritable, and it keeps no record of being wronged. It does not rejoice about injustice but rejoices whenever the truth wins out. Love never gives up, never loses faith, is always hopeful, and endures through every circumstance ... love will last forever!

Corinthians 13:4-8, Holy Bible

These few lines from the Holy Bible are so beautiful and perfect. As if all the explanation you need for your overwhelming feeling towards the loved one is there. Love is really an amazing thing. Perhaps it is the most amazing essence of being a human. All animals do it. All of you definitely have fallen in love at least once in your lifetime. The first time, you laid eyes on your dearly beloved, you suddenly started to feel the soothing breeze brushing against your skin, hear the sweet chirping of the birds, even the time seemed to have ceased for you two. For those of you who have recently fallen in love, you'd know even more clearly what I'm saying. As they say, there's nothing crazier than love. And it's really worth being crazy in love. Evolutionary speaking it works as a great motivator, as when you are in love more Oxygen rushes to the brain. And more Oxygen in the brain means more cerebral activity. But there's more to it than just increased O2 supply.

Love is a really complex neurobiological phenomenon, involving lust, attraction, trust, belief and pleasure activities within the brain. But let's keep it as much non-technical as possible and explore the chemistry of love as interestingly as making love.

What is Love?

Do we really want to know! As long as it keeps us up on our feet, who cares!!!

Every human being has his or her own perspective about love. A thousand people would define love in a thousand ways. Attachment, commitment, intimacy, passion, possessiveness, grief upon separation, and jealousy are but a few of the emotionally loaded terms used to describe the representation of love. In a lover's words: *"love is as vast as the cosmos, it is endless, limitless and unconditional"*. Although, if we start exploring the neurobiological mechanism of love with science, it becomes even more exciting and multi-dimensional. If you observe closely enough the astounding neurobiology of love itself can make your brain filled with O2.

This beautiful neurobiological mechanism critically involves various neurochemicals like oxytocin, vasopressin, dopamine, serotonin and endorphins. These neurochemicals make you love and feel loved.

The concept of love involves having an emotional bond to someone for whom one yearns, as well as having sensory stimulation that one desires. The word "love," however, derives etymologically from words meaning "desire," "yearning" and "satisfaction" and shares a common root with "libido" (sex drive).

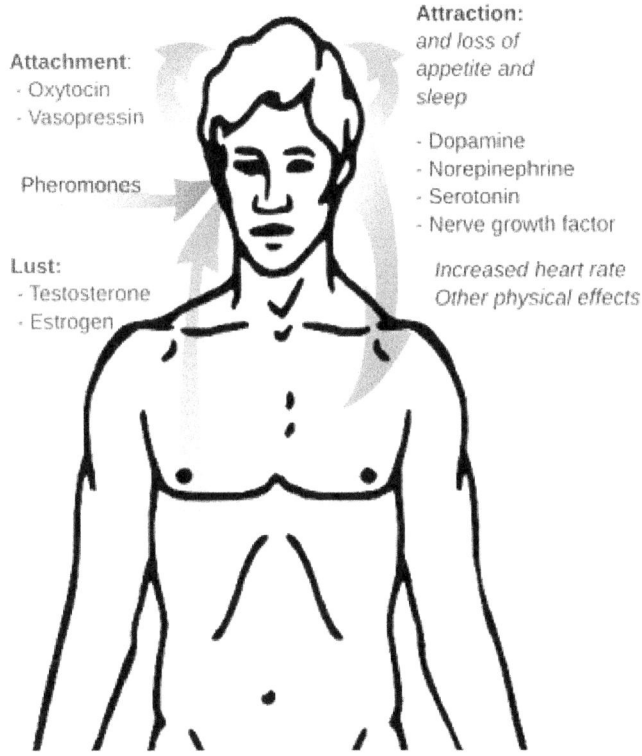

Figure 1.1 Neurobiology of Love

Thus, the psychological sense of love can be interpreted as referring to the satisfaction of a yearning, which may be associated with the obtaining of certain sensory stimulation. Love therefore possesses a close connection not only with reward and pleasure phenomena, but also with appetitive and addictive behaviors. This is the usual addiction of love, when you feel so much addicted to the person you love. Separation from the beloved one leads to withdrawal like symptoms, because your body literally keeps craving for the pleasure stimulation of being in love.

Love and its various emotional states and behaviors are rarely investigated by scientific means. In part, this may be due to the fact that love has always been the domain of poets, artists and philosophers. Although, for ages love has inspired many scientists in their extraordinary discoveries, it has certainly not been considered to be right within the scope of common experimental science, i.e., neurobiology research. Emotions and feelings such as attachment, couple and parental bonding, and love, neglected for centuries by the experimental sciences, have now come into the focus of neuroscientific investigation in order to elucidate the biological mechanisms and pathways of the most pleasing concept of humankind. So let's explore the ingredients of the love phenomenon.

Love begins with the stage of primitive lust and attraction. I'm saying primitive because at this very early stage there is really no difference between primitive man and modern man. The bodily characteristics of a person such as, how hot they are, poke the level of sex hormones (testosterone and estrogen), cortisol and pheromones. Lust is initiated at this stage through the physical attraction and flirting. This is an evolutionary behavior of mankind that biologically enables a human to find a healthy, fertile and perfect mate.

Following the cue of lust, the major attraction symptoms kick in, which are usually known as the symptoms of love, such as sweaty palms, tremors in the whole body, restlessness, loss of appetite and sleep, thumping heart, butterflies in the stomach etc. Such symptoms occur because the body is flooded with neurochemicals like Dopamine, Cortisol, Norepinephrine, and Phenylethylamine (PEA). Once this euphoria wears off, the ultimate and deepest stage of love prevails that is the attachment phenomenon. And the chemicals that make this possible are

Oxytocin, Vasopressin and Endorphins. As time goes by, the crazy love sensation diminishes and the feeling of closeness and attachment grows and prevails till the last breath of life.

Love is more than just a blissful Kiss. It is a really pleasurable experience, moreover it's the best natural motivator. Love turns on the reward center of your brain like a light bulb. Everything around you seems so beautiful. You feel an amazing bliss with your beloved partner. As a result deeply intriguing expressions of love come out of your mouth: *"You and I are forever"*, *"You drive me crazy like hell"*, *"I belong to you"*.

The intensely sensational and emotional state of love has inspired artists to create masterpieces and even scientists to come up with world-changing ideas like Erwin Schrodinger's Wave Equation and Stephen Hawking's Hawking Radiation. That's why we neuroscientists are so fascinated with the love phenomenon.

The exhilarating sensation of love cannot be tamed by any law of the sophisticated human civilization. When it overwhelms a person with amorous desires it does not care about whether the person is a philosopher, physicist or a neuroscientist. Take the Austrian Physicist Erwin Schrodinger for example. He is a legend of Quantum Mechanics. Responsible for major advancements in quantum mechanics, Schrodinger's name is now synonymous with what is known as the "Schrodinger's cat" paradox, a thought experiment in which a cat, trapped in a box with a breakable vial of poison, can be considered both alive and dead by the outside observer until the box is opened, illustrating the principle of superposition in quantum theory.

Figure 1.2 Obverse side of the old Austrian 1000 Schilling note with the Austrian physicist Erwin Schrodinger

In 1926 Schrodinger presented a theory explaining the spectrum of the hydrogen atom, likening its functioning to a wave equation, thus laying the foundation for wave mechanics. His ideas of wave mechanics were inspired by de Broglie's ideas, which he had first thought to be completely rubbish, until he was persuaded otherwise. Now comes the juicy part, in the words of the physicist Hermann Weyl, Schrodinger obtained his inspiration for wave mechanics while engaged in a *"late erotic outburst in life"*. He was what we can the greatest Casanova among all scientists. To be more specific, he had a wandering eye for young girls on the brink of adolescence. In the book Erwin Schrodinger and the Quantum Revolution, biographer John Gribbin portrays the man as: *"He was often in love – or he convinced himself that he was in love – and when he was in love, by and large life was good and his scientific creativity benefited."*

In 1920 he married Annemarie Berthel and despite all the stormy episodes in the relationship they remained together until his death. He had no children by his wife, but he had at least three illegitimate daughters.

One wonders whether Schrodinger's famous wave equation was a product of marital bliss. It was indeed a product of love, but there's a twist in the story. As I said earlier he was an amorous philanderer. His wave equation, the key equation of quantum physics emerged during one of his amorous adventures in late 1925, when he stayed at a hotel resort with a mistress while his wife remained in Zurich. The woman involved in this particular encounter has not been identified, but whoever she was, she deserved to be admired as the *"greatest muse of a great thinker"*.

Arnold Sommerfeld once referred to the Schrodinger's wave equation as: *"the most remarkable of all remarkable discoveries of the 20th century"*. When Max Planck held Schrodinger's second publication in his hands, he sent Schrodinger a post card saying:

"I am reading your communication in the way a curious child eagerly listens to the solution of a riddle with which he has struggled for a long time, and I rejoice over the beauties that my eye discovers, which I must study in much greater detail, however, in order to grasp them entirely."

With reference to the origin of the Schrodinger equation, the American Nobel laureate Richard Feynman noted:

"Where did we get that [Schrodinger's equation] from? Nowhere. It is not possible to derive from anything you know. It came out of the mind of

Schrodinger, invented in his struggle to find an understanding of the experimental observation of the real world."

Figure 1.3 Tombstone of Annemarie and Erwin Schrödinger (1887–1961) with the Schrodinger equation

So, the bottom-line is that the erotic adventure of Erwin Schrodinger led to the birth of the key equation of quantum physics. He won the Nobel Prize for Physics in 1933 for his 1926 introduction of the wave equation. He was a true genius of modern physics, but his innate biological urges didn't know the bounds of the society. He could not keep his hands of teenage girls. He had children by at least three of his mistresses, including a daughter by Hilde March, the wife of his colleague Arthur March. His wife Anny once said: *"it would be easier to live with a canary bird than with a racehorse, but I prefer the racehorse."* In the face of primitive urges, even the great physicist turned out to be helpless. That's how strong the early stage of lustful attraction can be.

Molecular signaling can make the experience of early love pleasant and stressful at the same time. Until you really express your feelings to the person, there remains a sort of uncertainty, which is kind of stressful. This uncertainty leads to a duel inside your mind and therefore increases the cortisol level. Once you spill the eggs, the stress hormone cortisol level goes down.

The very first phase of love gives rise to a crazy euphoric sensation. The passion of love creates the feeling of exhilarating happiness that is often unbearable and certainly indescribable. And the areas that are activated in response to romantic feelings are largely co-extensive with those brain regions that contain high concentrations of a neuro-modulator called dopamine that is associated with reward, desire, addiction and euphoric states. Like two other modulators, oxytocin and vasopressin that are linked to romantic love, dopamine is released by the hypothalamus, a structure located deep in the brain and functioning as a link between the nervous and endocrine systems.

These same regions become active when exogenous opioid drugs such as cocaine, which themselves induce states of euphoria, are ingested. Release of dopamine puts one in a "feel good" state, and dopamine seems to be intimately linked not only to the formation of relationships but also to sex, which consequently comes to be regarded as a rewarding exercise. An increase in dopamine is coupled to a decrease in another neuro-modulator, serotonin (5-HT or 5-hydroxytryptamine), which is linked to appetite and mood.

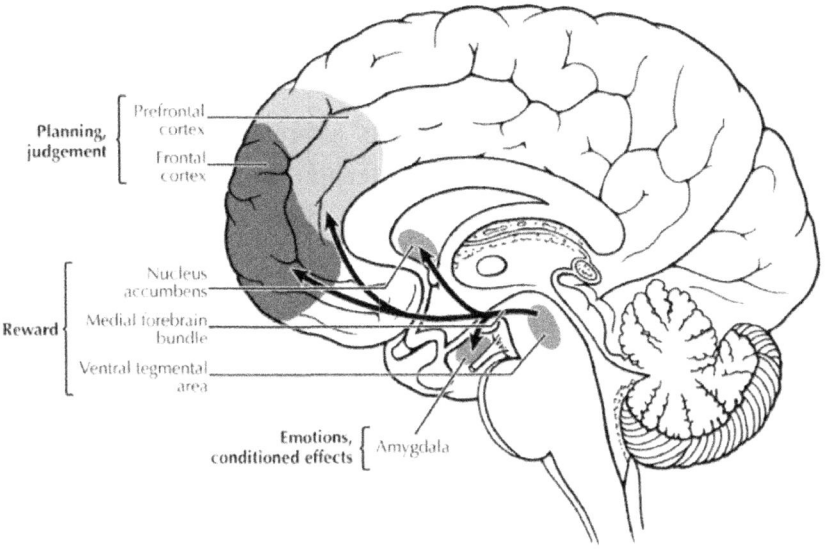

Figure 1.4 Brain Areas involved in pleasure/reward, judgment and emotional effects

Love, after all, is a kind of obsession and in its early stage commonly immobilizes thought and channels it in the direction of a single individual. At this early stage you cannot think of anything else but that apparently perfect special person. Different regions of the brain mostly analytical and logical ones like the prefrontal cortex become less active. That explains why people who are in love are not able to judge their partner's character honestly. Come on, let's be honest, we have all felt like this, when we first fell in love – as if we look at the special person "Through Rose-Colored Glasses".

Euphoria and suspension of judgment can lead to states that others might interpret as madness. It is this madness that poets and artists have celebrated.

Plato considered it in "Phaedrus" as a productive, desirable state. But this cognitive blindness towards your beloved one's dark side at the first stage of romance has an evolutionary importance. It helps you pass the mad love stage without being confused by any behavioral response from the other. At the beginning of a romantic relationship both the individuals pose to be anything but themselves. It's not until the later attachment and bonding stage when the two individuals start to open up to each other.

In Plato's Phaedrus, Socrates comments:

> *"the irrational desire that leads us toward the enjoyment of beauty and overpowers the judgment that directs us toward what is right, and that is victorious in leading us toward physical beauty when it is powerfully strengthened by the desires related to it, takes its name from this very strength and is called love"*.

There are no moral strictures in the early stage of love, for judgment in moral matters is suspended as well. After all, moral considerations play a secondary role, if they play one at all, with Anna Karenina, or Phedre, or Emma Bovary or Don Giovanni. And morality, too, has been associated with activity of the pre-frontal cortex. So it is really important that the analytical and judgmental pre-frontal cortex is partially switched off, in order to survive the early stage of intense and crazy sensation of lust, crush, infatuation, attraction and obsessiveness. The early stage of romantic love seems to correlate as well with another substance, nerve growth factor, which has been found to be elevated in those who have recently fallen in love compared to those who are not in love or who have stable, long-lasting, relationships. Moreover, the concentration of nerve growth factor appears to correlate significantly with the intensity of romantic feelings.

As time passes, the euphoria of love wears off and serotonin level gets back to normal. Now begins the most mature stage of true love, that is attachment and bonding through closeness and sexual intimacy. Here passionate love transforms into compassionate warm love. The ingredients of this stage are Oxytocin and Vasopressin. Alongside the feel good chemical Dopamine, these two neuro-modulators are the most prominent hormones in attachment and bonding. Both are produced by the hypothalamus and released and stored in the pituitary gland, to be discharged into the blood, especially during orgasm in both sexes and during child-birth and breast-feeding in females. In males, vasopressin has significant link with social behavior, in particular to aggression towards other males. The concentration of both neurochemicals increases during the phase of intense romantic attachment and pairing. The receptors for both are distributed in many parts of the brain stem which are activated during both romantic and parental love.

Parental love is another crucial ecstatic experience of the human species. Oxytocin, Vasopressin and Prolactin are the major chemicals of parental love and bonding. Oxytocin is one of nature's chief molecule for creating a mother. Roused by the high level of estrogen during pregnancy, the number of oxytocin receptors in the expecting mother's brain multiplies dramatically near the end of her pregnancy. This makes the new mother highly responsive to the presence of oxytocin. It is the key component to ecstatic and orgasmic childbirth. Oxytocin's first important surge occurs during labor. If a cesarean birth is necessary, allowing labor to occur first provides some of this bonding hormone surge and helps ensure a final burst of antibodies for the baby through the placenta. Pressure against the birth canal further heightens oxytocin levels in both mother and baby.

Vasopressin induces the protective nature in a father for his partner and child. During pregnancy testosterone level starts to drop in the father to be. As a result, it reduces the aggressiveness in the male. And prolactin level increases in the father which brings the paternal caregiving instincts. Elevated prolactin levels in both the nursing mother and the involved father cause some reduction in their testosterone levels, which in turn reduces their libidos. Therefore, thanks to Mother Nature's amazing craftsmanship, parental love becomes an ecstatic experience just like romantic love.

All that two lovers want, is to be one with the other in a moment of ecstasy. They want to be ONE in every way possible, being entangled in each other. Sexual union is the closest that humans can get towards achieving that unity. Through sexual intimacy, two human beings become one mind, body and soul wrapped up in the cocoon of their skin.

Love can literally transform a human being. It can make us do either heroic or evil deeds. It is the best inspiration ever. Artists have created various masterpieces inspired by love. William Shakespeare was one of those inspired ones. In "A Midsummer Night's Dream" he expresses romantic love as a dynamic power that always finds a way to overcome obstacles, is not straight-lined and alternates between phases of high and low current.

When you are in love (not in the early crazy love stage), you really are able to achieve greater things, due to increased cerebral activity and the feel good stimulant. While on the contrary, a break up leads to reduced brain activity. If a relationship comes to an end, it is usually experienced as an unpleasant event, with increased levels of stress hormones. Recent studies of brain activity patterns have shown increased activity in areas active during choices for uncertain rewards and delayed responses, reflecting a common feeling of uncertainty about the future. Rejected individuals showed a decreased activity in brain networks involved in the onset of major depression and also showed depressive symptoms, suggesting that the grieving period following a break up might be a major risk factor for clinical depression.

To understand the components of "perfect love" and the "perfect couple", Robert J. Steinberg of the Yale University postulated a theory called the "Triangular Theory of Love". This theory holds that love can be understood in terms of three components that together can be viewed as forming the vertices of a triangle. These three components are intimacy, passion and commitment.

The complete form of love or "consummate love" consists of all the three components in a balanced ratio, giving rise to the "perfect couple" in a long-term relationship. Such couple will continue to have fifteen or more years of great sex. They cannot imagine themselves happier over the long-term with anyone else. They overcome their difficulties gracefully, and each delights in the relationship with the other. A state that sounds desirable. Maintaining this state is highly dependent on a successful translation of the components into action.

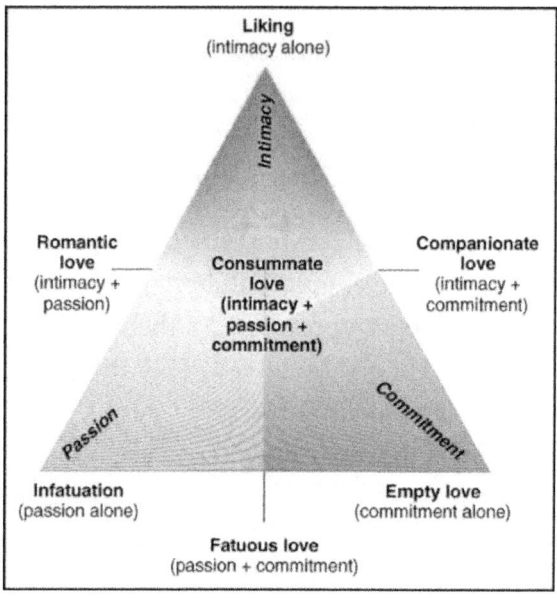

Figure 1.5 The Triangular Theory of Love

Consummate love may not be permanent. And remember not to cross the limit of your body's stress level while having blissful sex, because after that it'd become less pleasuring and more stressful, as then the release of dopamine reduces and the level of stress hormone cortisol rises.

A fulfilling long-term relationship is not accomplished by just finding "the one". It is rather a co-operation between two passionate and highly motivated partners working together, figuring out every single situation holding hands. If there is trust at the root of the relationship, if the partners make an effort to keep it interesting, if difficulties are handled tactfully and if you can appreciate every single deed of your partner no matter how insignificant it is, the flames of love would never burn out and your love can truly "live happily ever after".

Special Note:

A special tip for those in love. Chocolate is a wonderful romantic gift for your loved one as it contains Phenylethylamine (PEA), the love molecule which induces euphoria and pleasure in the brain.

Chapter 2

Connectivity of Minds - Think of The Friend and The Friend Appears

Most of you have often experienced the specific phrase from the title of this chapter: think of the friend, and the friend appears, or rather think of the devil and the devil appears. While you are thinking of your friend, suddenly quite out of the blue your smartphone dings and you see that the friend is calling or has sent you a text. And then you keep asking yourself over and over again, how is it even possible!!! Are you some kind of "psychic" or something!!!

Well that's what this chapter is going to elaborate on. Such a telepathic ability is one of the many marvels of human brain. It's a kind of Extrasensory Perception (ESP) which is an intricate neurological characteristic of every human brain. It is the product of millions of years long evolution. Earth-Brain Bondage (which is discussed in chapter 6) and your love and affection towards the friend make this extraordinary cognitive skill possible. In technical term it is called as "psi phenomenon". Throughout history there have been many cases of subjective experiences of telepathy or clairvoyance (T-C) involving reminiscence, death, sickness or crises of friends, couples and relatives. When T-C experience occur, they usually involve extreme emotional bondage between two human beings.

Human brain is a living transceiver, which is able to catch the emotional signals of another fellow brain. But how exactly the cerebral signals reach another brain irrespective of the distance in between. To explain this phenomenon, we have been developing hypothesis that ELF (extremely low frequency) electromagnetic fields are associated with T-C experiences. The ELF hypothesis is actually a generic label. In fact there are two most fundamental ELF fields that count for our extraordinary telepathic abilities, they are the Schumann's resonances (SR) and Geomagnetic pulsations. Since the time when human brain evolved, it has been immersed in these fields all along. Features of the Schumann resonance and geomagnetic field have been explained in chapter 6. The SR is responsible for relentless connectivity between two emotionally related minds or more specifically two cerebral biomagnetic fields. Geomagnetically quieter days allow the SR field to make the transference between two human brains more intense than the days with geomagnetic disturbances.

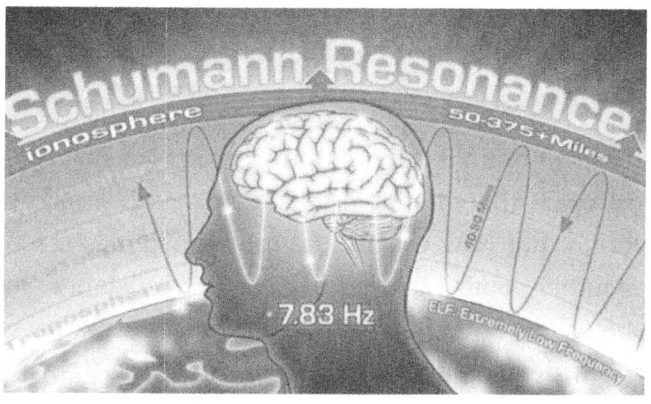

Figure 2.1 Schumann Resonance and Human Cerebral Field

In a study done in 1980s, it clearly shows that spontaneous subjective T-C experiences are more likely to occur on days when the geomagnetic activity is lower (quieter) than on days before or after the experiences. This pattern was statistically significant and was evident in the experiences that occurred between 1920 and 1967. So you can see that you are not the only person who may be experiencing such telepathic abilities. The strongest connectivity of minds is seen in mother-child relationship and in couples. For example, a mother is often able to feel the crises of her child who might be thousands of miles away. The same happens to couples who are deeply in love. If you are in love and you think that you can literally read your partner's mind before he or she even expresses something, then you really are a psychic with increased cerebral activity. With the assistance of the surrounding environmental electromagnetic fields the minds on planet earth remain connected forever.

But how exactly is it possible for a human mind to interact with another, through the carrier waves of environmental electromagnetic field! You will find in the later chapter Earth-Brain Bondage, human brain is able to feel, whatever our mother earth feels. These two systems share remarkably similar physical characteristics such as dynamic amplitudes and intrinsic frequencies. Therefore the possibility for direct interaction emerges. From this we can deduce that impedances and reluctances match or shared resonance occurs between the electromagnetic energies. The thin shell between the ionosphere and earth generates continuous harmonics of frequencies from a fundamental of about 7 to 8 Hz that is caused by global lightning which occurs between 40 to 100 times per second (40 to 100 Hz). Those frequencies are the Schumann resonances.

The bulk velocity of neuronal activity around the human cerebral cortices caused by the discharge of action potentials within this thin shell of tissue generates a resonance with a fundamental frequency between 7 to 8 Hz. On the bases of the average durations of the travelling waves over the cerebral cortices the repetition rate and phase velocities are in the order of 40 to 80 times per second (40 to 80 Hz). The current density around the annulus of an axon associated with a single action potential is equivalent to about 10^5 A·m^{-1} or the value associated with a single lightning discharge. For both cerebral and earth-ionosphere phenomena the average potential difference for these time-varying processes is in the range of 0.5 mV·m^{-1}. The magnetic field component is in the order of 2 pT (10^{-12} T). The ratio of this voltage gradient to the magnetic field intensity is effectively the velocity of light, 300,000 km·s^{-1}.

Such quantitative evidence implies that the electric and magnetic fields of the Schumann resonances and those equivalent frequencies generated by the human cerebral cortices can interact persistently in a state of geophysical equilibrium or to speak simply, the communication occurs persistently in geomagnetically quieter days.

Now let's look into the mathematical aspect of the interaction between human mind and our planet. If you are not really into mathematics, you can just skip this part. A recent discussion of the Schumann Resonance characteristics summarizes that, the harmonics or modes of the Schumann resonances peak around 7.8 Hz, 14.1 Hz, 20.3 Hz, 26.4 Hz, and 32.5 Hz. This serial shift of about 6 Hz is consistent with the relation of:

$$[\sqrt{(n(n+1))}] \cdot [(v \cdot (2\pi r)^{-1})]$$

where n are serial integers ≥ 1, v is the velocity of light in the medium (which is effectively c) and r is the radius of the earth. The first component of the relation when "n" is substituted as a quantum number is also employed to calculate the magnitude of the orbital angular momentum. When magnetic moments are expressed as Bohr magnetons the electron state is associated with a magnetic moment equal to $\sqrt{(n(n+1))}$. Changes in electron states or different shells are associated with emissions of photons. The peak values are not precise and can vary by ±0.2 Hz depending upon ionosphere-earth conditions, time of day, season, influx of protons from solar events, pre-earthquake conditions, and yet to be identified sources. According to Nickolaenko and Hayakawa monthly variations in the first mode (7.8 Hz) range between 7.8 and 8.0 Hz. The diurnal variation in frequency shift has been attributed to drifts in global lightning or alterations in ionospheric height. Peak frequencies and amplitudes occur in May whereas minima occur in October-November. The peak to peak modulations are about ±25 % of the median value. The optimal metaphor is that every lighting strike of the approximately 40 to 100 per second between the ionosphere and ground is an expanding wave that moves until it ultimately interfaces with itself on the earth's spherical surface. The resulting interface elicits a return wave that arrives at the original source within about 125 ms after the initiation. The approximate equivalent frequency is 8 Hz.

The human cerebrum (1.35 kg) is an ellipsoid aggregate of matter that occupies about $1.3 \cdot 10^{-3}$ m^3. The three-dimensional metrics are: length (155-190 mm), width (131-141 mm), depth (108-117 mm). The cerebral cortices are approximately 1 to 4 mm thick but occupy almost 40 % of the cerebral volume with an average value of 490 cc. The surface area of the human brain is not smooth but exhibits convexities (gyri) and concavities (sulci). Two-thirds of the surface is buried within the sulci. Mathematical modeling indicates that this topological surface is similar to that of a flat surface "wrinkled" into the third dimension.

The primary source of the electromagnetic activity measured from the scalp emerges from the cerebral cortices because of the parallel arrangement of the dendrite-soma-axon orientations perpendicular to the surface for most of the approximately 20 billion neurons. The resulting steady state potential between the cerebral surface and a relative reference such as the lateral ventricle ranges between 10 and 20 mV. There is almost a linear correlation between the emergence of neuronal processes, and the magnitude of that potential. Superimposed upon the steady potential are fluctuating voltages that define the electroencephalogram (EEG). Most of the discernable frequencies occur within the ELF range of 1 and 100 Hz. However, fast frequencies up to 300 to 400 Hz, approaching the absolute refractory period of an axon, have been measured in epileptic brains.

The amplitudes of the scalp EEG range between 10 and 100 µV. In comparison corticographic discharges exhibit amplitudes between 0.5 and 1.5 mV. The most prominent dynamic pattern is the alpha rhythm (8 to 13 Hz) which during wakefulness is more evident over the posterior cerebral space with amplitudes in most (68 %) of the population between 20 and 60 µV.

Theta rhythms (4-8 Hz) have been considered "intermediate waves" that are significantly involved with processes associated with infancy and childhood as well as drowsiness and specific stages of sleep. The power of theta activity is evident even within the third decade (25 to 30 years of age) when the EEG parameters asymptote. This inflection is sometimes considered an index of cerebral maturation.

The other classical frequency bands associated with the human EEG are delta (1-4 Hz), which are the highest amplitude time-variations associated with Stage IV sleep, beta (13- 30 Hz) and gamma (30 to 50 Hz) patterns. The ranges are effectively arbitrary and related to EEG features associated with particular behaviors or power densities.

So, the fundamental brainwave EEG happens to fall within the basic harmonics of the Schumann Resonance peak around 7.8 Hz, 14.1 Hz, 20.3 Hz, 26.4 Hz, and 32.5 Hz. The typical strength of the electric field (SR peak harmonics) component is in the order of $mV \cdot m^{-1}$ while the magnetic field component is ~1 $pT \cdot Hz^{-1/2}$. Compared to the troughs of about 1 pT the peak intensities at the first and second harmonics are ~3 pT which decline to ~1 pT around 20.3 Hz and <0.5 pT at higher harmonics. The fundamental frequency and intensity increase by ~.04 to .12 Hz and 0.11 to 0.41 pT, respectively during strong proton events (solar storms). The coefficients of this magnitude are slightly but significantly different along the X,Y,Z axes of propagation. Although these values appear minute, their potential can be realized by the magnetic energy from these values, described by:

$$J = B^2(2\mu)^{-1} \, m^3$$

where B is the magnetic field strength, μ is magnetic permeability ($4\pi \cdot 10^{-7}$ N·A^{-2}), and m^3 is volume of the human cerebrum (~$1.3 \cdot 10^{-3}$ m^3). The solution is within the range of ~10^{-20} J. Just like the Unified Field Theory in modern physics, many of us neuroscientists are in an effort to come up with a Unified Theory for Neuroscience. The neuroquantum unit 10^{-20} J is the first step towards the Unified Theory. This quantum unit of energy is associated with the effect of the $1.2 \cdot 10^{-1}$ V net change during an action potential of a neuron upon a unit charge ($1.6 \cdot 10^{-19}$ A·s). That the activity of one neuron can affect the state of the entire cortical manifold of the brain has been shown experimentally.

In simple words, this single quantum of energy (10^{-20} J) from the Cerebral Electromagnetic field of thoughts and emotions would have the potential to disperse through the Schumann resonance ELF and reach another Cerebral Field thousands of miles away. In geomagnetically quieter days the environmental electromagnetic field carries the potential information of the specific thought or emotional response to the other brain and shares it with the cerebral field. But when there is disturbance in the geomagnetic field the inter-cerebral signals get disrupted, just like during thunderstorm sometimes your cell network gets lost. Geomagnetic disturbance occurs due to different terrestrial and cosmological reasons, like solar storm (Coronal Mass Ejection CME), cosmic rays, lunar gravity interaction during full moon.

The cerebral fields of two emotionally involved human brains attain tuned brain wave EEG through the persistent Schumann resonance field which is in touch with both the brains at any specific moment, which means the more you get emotionally involved with your loved one, the more you'll be able to read each other's minds. Once the minds are truly engaged, the next step would be to feel the other's physiological state of being. When you love someone deeply, every single cell in your body starts reacting when that special person is in distress.

Chapter 3

Science of Empathy & Learning - The Mirror Neurons

Empathy is one of the various precious human characteristics that allows the humans as a species to connect with each other or even other creatures of earth. It is a bonding-mechanism between the living beings on our planet. Thanks to this amazing quality, you are even sometimes able to feel the pain and emotions of others. But how does this really work scientifically! Well… behind the curtain it's the work of another neurological wonder called the Mirror Neurons.

Most of you readers can already make sense out of the name itself. It works as a neurological mirror which reflects the emotional response through you from the person in front. Which means, when you see a couple hugging and kissing in front of you, don't be amazed if you start feeling the warmth of your beloved one's passionate embrace. As humans, we are evolutionarily designed to empathize with the other and share his/her grief or happiness even if that person is a complete stranger.

These marvelous cells of the human brain have literally shaped our civilization and these tiny wonders within our brain, make us aware of what it's like to be human. They are also the reason why you start yawning or giggling merely by watching another person do the same. They play the key role in a child's brain, while it is learning its mother tongue along with other cultural and sociological tactics. And as for adults, mirror neurons are your wingmen who help you in learning new skills. In the 1980s and 1990s, a few researchers Giacomo Rizzolatti, Giuseppe Di Pellegrino, Luciano Fadiga, Leonardo Fogassi, and Vittorio Gallese at the University of Parma, Italy discovered the mirror neurons in macaque monkey. The discovery was initially sent to Nature but was rejected at that time due to "lack of general interest", only to be accepted widely in a few years.

Mirror neurons were first found in various regions of the monkey brain (Macaca nemestrina and Macaca mulatta). So far scientists have observed the wonderful neurons in areas like the ventral premotor cortex (vPMC), inferior parietal lobe (IPL), primary motor cortex and dorsal premotor cortex (dPMC). Originally it was discovered, that the mirror neurons (MN) discharge both when the monkey does a particular action and when it observes another individual (monkey or human) doing a similar action. The name itself implies the significant feature of MNs. This specific feature of 'mirroring' or more specifically 'imitating' has been evolutionarily crucial in shaping the modern human civilization.

Figure 3.1 Mirror Neuron activity in the monkey brain, when it grasps the banana as well as when it observes another individual grasping it

Neurophysiological experiments demonstrate that when humans observe an action done by another individual their motor cortex becomes active even without the presence of any motor activity. On this matter the first evidence was provided in the 1950s by Gastaut and his coworkers. They recorded significant changes (desynchronization) in an EEG rhythm (the mu rhythm) of humans both during active movements of studied subjects and when the subjects observed actions done by others.

Later many other researchers replicated the experiment and confirmed the results. The desynchronization while observing an action carried out by others includes rhythms originating from the cortex inside the central sulcus. Transcranial magnetic stimulation (TMS) studies provide us even more direct evidence to the existence of mirroring properties in the motor system of humans. TMS is a non-invasive technique for electrical stimulation of the nervous system. When TMS is applied to the motor cortex, at appropriate stimulation intensity, motor-evoked potentials (MEPs) can be recorded from contralateral extremity muscles. By modulating the amplitude of the MEPs through behavioral stimulation, it is possible to assess the significant effects of various experimental conditions. This way we can study the mirror neuron system in humans.

In the year 1995 Fadiga et al recorded MEPs, elicited by stimulation of the left motor cortex, from the right hand and arm muscles in volunteers required to observe an experimenter grasping objects (transitive hand actions) or performing meaningless arm gestures (intransitive arm movements). The results showed that the observation of both transitive and intransitive actions made a huge impact over the MEPs and increased the recorded MEPs. The increase concerned selectively those muscles that the participants use for carrying out the exact observed actions. Amplification of the MEPs while observing actions done by the others may result from the activation of the primary motor cortex owing to mirror activity of the premotor areas.

Later in the year 2000 another study by Strafella & Paus came up with support for this cortical hypothesis. By using a double-pulse TMS technique, they demonstrated that the duration of intracortical recurrent inhibition, occurring during the observation of an action, closely corresponds to that occurring during the execution of that specific action.

Another question rose from these studies, "Does the observation of actions done by others influence the spinal cord excitability?" In 2001 Baldissera and colleagues investigated the issue and discovered that there is an inhibitory mechanism in the spinal cord that prevents the execution of an observed action and leaves the motor cortex free to react to that action without the risk of any kind of movement generation. Later many other neuroscientists showed that the motor cortical excitability faithfully follows the grasping movement phases of the observed action, or to say simply, your motor cortex area of the brain lights up whenever you see another person carrying out an action as well as when you carry it out yourself.

In conclusion, TMS studies indicate that a mirror-neuron system exists in humans and that it possesses important properties not observed in monkeys. Intransitive meaningless movements produce mirror neuron system activation in humans but not in monkeys. These properties of the human mirror neuron system (MNS) play an important role in determining the humans' capacity to imitate others' action as well as empathizing with them.

The Mirror Neuron System has vast impact over a person's social and behavioral skills throughout the lifetime. It allows us to be human and understand another human being and even other species for that matter. When you see a person get beaten up in the park, you suddenly start to feel his agony. The same happens when you see a street dog getting hurt. Humans are biologically designed to truly understand another creature's pain, happiness and desires, as if it is our own pain, happiness and desires. Feeling other's emotions and imitation learning are the most influential features of the mirror neurons.

Imitation is a great and the best mechanism of learning new skills in children and adult alike, although it is the most widely used form of learning during development, offering the acquisition of many skills without the time-consuming process of trial-and-error learning. Imitation is also central to the development of fundamental social skills such as reading facial and other body gestures and for understanding the goals, intentions and desires of other people.

Many of us neuroscientists believe that there's a strong possibility that malfunction in the imitation learning mechanism or more specifically in the MNS may underlie various cognitive disorders, especially Autism.

During the developmental age of a child it learns language and different social skills. At this crucial stage of a person's lifetime, dysfunction in the mirror neurons might be one of the core deficits of socially isolating disorders such as Autism. A recent study by Iacoboni and colleagues highlights the importance of mirror neurons and their role in the development of autism spectrum disorder (ASD).

ASD is a pervasive developmental disorder characterized by impaired social interactions. Iacoboni's team used functional magnetic resonance imaging (fMRI) to investigate neural activity of 10 high-functioning children with ASD and 10 normally developing children as they observed and imitated facial emotional expressions. Although both groups performed the tasks equally well, children with autism showed reduced mirror neuron activity, particularly in the area of the inferior frontal gyrus. Moreover, the degree of reduction in mirror neuron activity in the children with autism correlated with the severity of their symptoms. These results indicate that a healthy mirror neuron system is crucial for normal social development. If you have broken mirrors or deficits in mirror neurons, you likely end up having social problems, as patients with autism do.

What about learning new skills in adults! Well... say you have moved as a professional to Paris, the most romantic city in the world with your beloved one. Although, you will be able to communicate with locals of Paris in English, but in order to look inside a Frenchman's inner self and understand his intentions, you'd require to learn the language of Paris' heart. In that case the neural basis of learning the sexy French, would be to listen to a lot of Frenchmen speaking French. While your brain observes others speaking a foreign language it'd start firing in the Mirror Neuron Network and try to make sense out of it, by making you absorbed in the language either consciously or subconsciously. The MNS would make you feel as if you are speaking the language in your mind. Eventually as you keep on listening to French along with a little literary assistance, the language centers of the brain; the Broca's area and Wernicke's area would start forming new neural network responsible for communicating in French.

Likewise, this imitation learning mechanism is responsible for enabling you to learn any new skills, like music, dance, mathematics etc. The more you observe and practice a specific skill, the MNS would fire more and the responsible area of the brain forms new neural pattern. This way you become better and better at a specific skill.

Humans are intensely social creatures. We share this feature with many other species. A complexity and sophistication that we do not observe among ants, bees or wolves, however, characteristically define the social life of primates. This complexity and sophistication is epitomized at its highest level by the social rules. Living in a complex society requires individuals to develop cognitive skills enabling them to cope with other individuals' emotions, intentions and actions, by recognizing them, understanding them, and reacting appropriately to them. In a study, Singer and her colleagues used functional magnetic resonance imaging (fMRI) to measure brain activity in volunteers who observed others receiving painful stimulation to their hands. As expected, mere observation of another's pain produced increased activation in the pain center of the observer's brain, including the insula and anterior cingulate cortex.

In the earlier study of Singer et al. (2004), people who scored higher on standard empathy scales had higher activity in these brain areas. It thus appears that more empathetic people have more active mirror neuron systems for appreciating the pain of others. The major manipulation of the Singer et al. (2006) study was that the people who received painful stimulation had previously engaged in a game where some had behaved fairly and others unfairly. Men, and to a lesser extent women, showed much less pain center activation in the brain for those sufferers who had acted unfairly. Moreover, men but not women showed greater activation in the reward center or pleasure center of the brain, the nucleus accumbens when observing unfair people being punished. Thus men more than women took pleasure in the pain of wrongdoers.

We literally live by emotions. Knowingly or unknowingly we always get triggered by another person's intentions and feelings. Mirror Neuron

System makes it possible for us to empathize with a fellow human being. That's why movie stars are so good at making the viewer burst into tears over a scene. When the actors put all their excellence and effort in making the characters real, the viewer brain starts to feel the character's suffering. You start feeling as if you are the character in the movie yourself. Which means, while watching Casablanca, for a few hours you become Humphrey Bogart or Ingrid Bergman yourself. Hence, the Mirror Neuron System acts as a bonding agent among humans through the mechanism of empathy.

Chapter 4

Meaning of Dreams

Dreams are the gateway towards one's soul. Many consider them as a path to the other world, the supernatural kind. And even, throughout human history many people have received vision of God or commands from the Supreme Soul in dreams. Many even see their dead son, wife, husband or other dead relatives in their dreams. Even great scientists have perceived scientific revelation in their dreams. As if the cosmic record of information opens up to them through their dreams.

But, are there really any scientific basis to all these experiences! Actually yes. All these can be explained without involving the supernatural component. Dream-Interpretation is one of psychology's most celebrated matters of interest.

Basically dreams are brain's own mechanism of getting rid of unnecessary information throughout the daytime. It's a process of reverse memory. Often they also produce a virtual world of wish-fulfillment, where the most intense wish of yours comes true. This is just another way to keep you satisfied with the virtual simulation of the things you crave for. The world of dreams is really a mysterious and wonderful one. Let's explore the world.

Dreaming is a byproduct of REM (Rapid Eye Movement) sleep, while the brain works on the daily calibration. But first let's answer the question what a dream really is!

Well; dream refers to the subjective conscious experiences we have during sleep. We neuroscientists define dream as a subjective experience during sleep, consisting of complex and organized images that show temporal progression of visual imageries. Dreaming is a universal feature of human experience.

Why do we have vivid, intense, and eventful experiences while we are completely unaware of the world that physically surrounds us? Couldn't we just as well pass the night completely unconscious? Not really. The function of dreaming has remained to be a persistent mystery for a long time, but with the discovery of various active and inactive brain regions during REM dream state we are finding out the impact of dreams over the awake state of mind.

Cognitive neuroscientists and psychologists throughout the human history have put forward various theories on the function for dreaming. Dream consciousness is perceived as some kind of noise generated by the sleeping brain as it fulfills various neurophysiological functions during REM sleep. But don't even for a second think that this noise is completely meaningless. This is one of the most important meaningful noises in the world of biology.

In many cultures dreams have been hailed as messages from the gods and dismissed as random hallucinations. The pendulum of popular opinion has swung from one extreme to the other throughout recorded history and between cultures and camps, with scientists, psychologists, sages, and philosophers all weighing in. Aristotle, for one, believed dreams were formed by the dreamer's impaired mind, and Plato argued that dreams represent a frightening breakdown of reason.

In the Victorian era, some scientists posited that dreaming was pathological. But rather than placing dreaming on a mystical pedestal, or looking at dreaming as a deficient form of consciousness, it is highly instructive to look at dreaming as an alternative form of consciousness and a different way of thinking.

Though Plato and Aristotle could not have proven it, today we know that the dreaming brain is, in a sense, differently abled. At least two important regions, the dorsolateral prefrontal cortex (DLPFC) and the precuneus in the parietal lobe, are deactivated during rapid eye movement (REM) sleep, the period when most dreaming takes place. Because of this, we lack the ability to fully exercise our short-term memory when we dream, both within the dream and upon awakening. This helps explain breaks in continuity during the dream and why it is difficult to recall dreams on waking. Also, we are unable to locate our physical body in space when asleep, which is why the dreamer does not realize her or his body is at home in bed during nocturnal adventures in familiar or fantastical landscapes. Making decisions or directing will is likewise difficult while dreaming as the regions responsible for those cerebral activities are inactive during the REM sleep.

My own fascination with dreams began after studying a renowned work of Sigmund Freud called Interpreting Dreams. He was one of the early frontiers of psychology who dared to step into the mysterious arena of dream-analysis. He is known as the father of psychoanalysis. But Freud had a major flaw in his works. He presumed that all behavioral cerebral activities have something to do with libido.

Just imagine, according to Freud all your activities are connected to sexuality. Come on… let's be honest. It doesn't take a neuroscientist to figure out that it's not true. Sorry Freud, your idea of sex drive driving all brain functions doesn't hold water. But Freud did have some really intriguing functional theories about dreams. One of the most important of Freudean ideas about dreaming was wish-fulfillment. Sigmund Freud and Carl Jung mentioned in their works that *"dreams are the royal road to the unconscious"* and *"the most readily accessible expression of the unconscious"*. I'll elaborate on that in a while.

But first let's look into the neurobiology of dreaming. To understand dreams first you need to understand various stages of sleep in which dreams occur. Previously experts divided sleep into five different stages. Fairly recently stage 3 and 4 have been combined, so in your sleep you usually pass through four stages: 1, 2, 3 and REM sleep. These stages progress cyclically from 1 through REM then begin again with stage 1. Stages 1 to 3 collectively is known as Non-REM sleep. A complete sleep cycle takes an average of 90 to 110 minutes.

Now the question may rise in some of your minds is that what are REM and NREM sleep? Most of you may have already heard about REM sleep and Non-REM sleep. Rapid Eye Movement sleep is a very crucial stage of sleep characterized by the rapid and random movement of the eyes.

In 1953, Aserinsky and Kleitman discovered human rapid eye movement (REM) sleep and documented that dream reports were obtained most frequently when subjects were awakened from REM sleep. People with suspected sleep disorders lack REM sleep. Mostly dreaming occurs in REM sleep, but here is a long-standing controversy surrounding the existence of dream experiences during non-rapid eye movement (NREM) sleep. Dream reports from NREM sleep were less remarkable in quantity, vividness and emotion than those from REM sleep.

NREM sleep dreams are more frequent during the morning hours when the occurrences of REM sleep are highest. 80% of your sleep is NREM sleep while the rest 20% is REM sleep. There is little or no eye movement during NREM sleep. Each stage of NREM sleep lasts around 5-15 minutes.

Stage 1: Your eyes are closed, but it's easy to wake you up.

Stage 2: You are in light sleep. Your heart rate slows and your body temperature drops. Your body is getting ready for deep sleep.

Stages 3: This is the deep sleep stage. It's harder to rouse you during this stage, and if someone wakes you up, you would feel disoriented for a few minutes.

During the deep stage of NREM sleep, the body repairs and regrows tissues, builds bone and muscle, and strengthens the immune system. As you get older, you sleep more lightly and get less deep sleep. Aging is also linked to shorter time spans of sleep, although studies show you still need as much sleep as when you were younger. Which means that even though the sleep gets lighter with age, try to get as much sleep as you used to have during your youth in order to wake up rejuvenated.

REM sleep stage is the most crucial stage for dreaming. Newborns spend around half of their sleep in REM stage, which helps in the development of the baby brain. And studies have shown that REM sleep is very much important in learning new skills at all ages. One study found that REM sleep effects learning of certain mental skills. People taught a skill and then deprived of NREM sleep could recall what they had learned after sleeping, while people deprived of REM sleep could not.

Going deeper in the mysterious and captivating world of dreams we have found various regions of the brain that are active and inactive during REM sleep. While a great deal of our brain remains active in REM sleep, regions related to executive functions such as rational thought, linear logic and episodic memory, as well as primary sensory and motor functions remain relatively inactive or just simply asleep. Just imagine, all the regions that enable you to make sense of various events, are literally turned off. Then how can it be possible for the events or progression of imageries to be logical or real!!! All the imageries perceived during REM sleep are merely the by-products of random neural firings, while the brain turns on the calibration in order to be ready and refreshed for the next day activities.

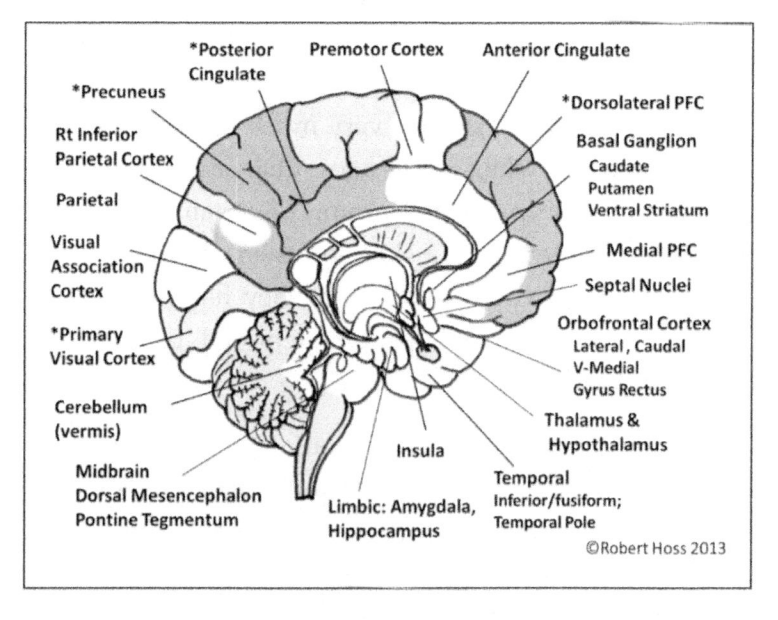

*Figure 4.1 Relatively active (white) and *inactive (dark gray) centers of the brain in REM sleep. Derived from neuroimaging studies (Maquet, Braun, Nofzinger et. al. in Hobson 2003)*

A number of active regions appear to be involved in the perception of the dream. Although the primary visual cortex and much of the parietal cortex remain inactive, activity is heightened in the visual association cortex, which processes imagery associations, and the right inferior parietal region, which organizes the imagery into a visual space.

Other sensory dream experiences may be due to internally stimulated activity in the vermis cerebellum and other motor and sensory regions as well as activity in the temporal areas involved in facial recognition, auditory processing and episodic recall.

Activity in the visual association regions of the cortex gives rise to the picture-metaphor nature of dream imagery, i.e. picturing the emotional, memory and conceptual associations.

This explains the visual imagery you have of your dearest one in your dreams. Here Freud's theory of wish-fulfillment proves its mettle. As an evolutionary advantage, your brain literally fulfills your most intimate wishes in your dreams and makes a virtual reality out of it, which doesn't seem any different than the real thing.

While you dream of your beloved one whether it is your wife, husband, lover, son or daughter, in your dreams you truly feel the oneness with them. In many cases it may not be a person people dream about, rather an object they wish to have. So if you crave for something or someone strongly enough, your dreams will give you the satisfaction of having them in a virtual reality, if not in real.

The majority of the brain centers which are active during REM dreaming are those which process mental material either unconsciously or prior to their actions becoming conscious. Dreams therefore provide us with valuable access to the unconscious. This proves that Freud and Jung were right about dream being the royal road to the unconscious. There are a number of cognitive centers in the frontal regions of the brain that are highly active in REM sleep. This suggests that the dreaming brain is capable of problem resolution and creative insight. For scientists who are truly absorbed in their work, even their unconscious mind keeps solving problems when they are having REM sleep. As a result, they often find solutions related to their work which they usually cannot find while being awake.

The great mathematician Srinivasa Ramanujan had visions of many of his mathematical equations in dreams. Most of the time in his dreams he used to see that a hand is writing various strange equations on a wall. Suddenly he used to wake up and start working on the dream imageries of those equations.

Figure 4.2 The Great Autodidact Mathematician Srinivasa Ramanujan

The creative problem solving history of dreams is well documented by Barrett (2001, 2007) who researched the many inventions and artistic creations arising from dreams. She describes dreaming as *"thinking in different biochemical state"*. Her research shows that anything may get solved during dreaming, particularly if the problem involves visualization or where the solution lies in "thinking outside the box".

But in most of the dreams people have, they see the metaphoric representation of their dynamic life situations. The metaphor of dreams becomes more obvious and eloquent when the dreamer tries to translate the dreams in his own words. The dream finds its own way of interpretation when it gets access to the dreamer's language center of the brain.

Let me give you an example; a man who had become frustrated and miserable at work but was not sure why, describes his dream:

"I had a frightening dream where I was being chased away by a big buffalo with a little buffalo following it."

When he was asked what a big buffalo does, he said:

"he is huge and powerful, when he wants you to go, you go,"

which he recognized as describing his boss. Then he was asked about the little buffalo. He said,

"He is a little pipsqueak that follows the big one around -- just like that little pipsqueak at work!"

The dream revealed that the source of his restlessness was not only the actions of his boss but the relationship that this little pipsqueak of a co-worker had with his boss. Here the metaphors aptly pictured the emotional similarities between the big buffalo and his boss, and the little buffalo and the 'little pipsqueak' at work.

Dreams are intensely influenced by emotions. The limbic areas, in particularly the amygdala, is highly active during the REM dream state. Dream creates imageries based on your emotional inclination. Emotion doesn't arise from the dreams, rather your deepest emotion orchestrates your dream. Many times your emotional feel-good state or frustrated state is woven into dreams with the use of picture-metaphors. A woman, who typically felt in control of the events in her life, suddenly learned from the doctor that her husband was terminally ill and she could do nothing about it. That night she dreamed,

> *"I was locked in a car with no steering wheel and no door handles or window controls. I was rolling backward down a steep hill, and there was no way of stopping it, or getting out of it. I woke up in a panic."*

Here the feeling of being totally unable to control the situation was pictured with all of its emotional intensity and vividness.

So as you can see, dreams are not just meaningless random neural firings of the brain. They really are the reflection of your unconscious mind. Basically we humans are emotional species. And this emotional trait even gets reflected in the dreams. High activity in the amygdala and associated limbic system has led us to conclude that dreams selectively process emotional memories through the interplay between the cortex and the limbic system.

So for now we can say that the amygdala actually orchestrates the dream activity. The virtual simulation of all emotional responses in your dreams in fact makes the brain gain more control over real life situations. Dreams have the potential to enable humans to handle even stressful circumstances smoothly.

Freud suggested that bad dreams let the brain learn to gain control over emotions resulting from distressing experiences. Studies have shown that dreaming metaphorically represents projections of emotional expectation and in many cases lowers stress levels (due to a massive reduction in stress producing neurotransmitters norepinephrine in forebrain centers including the amygdala).

Likewise nightmares are extreme form of emotional processing in dreams. But remember, people usually term every bad dream as a nightmare, while scientifically a dream is not a nightmare unless it is extremely upsetting, containing overwhelming anxiety, apprehension and fear. Nightmares can have a number of causes including heavy emotional stress, severe threat to self or self-image, unresolved or extreme trauma (PTSD), psychological problems, the influence of certain drugs, emerging medical problem requiring attention, or sleep disorders affecting REM/NREM sleep balance. Resolving the underlying cause can lead to the freedom from nightmares.

But not always dreams are so obvious and easily interpretable. Sometimes dreams are exquisitely bizarre that really just don't make any sense. In that case, you don't need to be confused, rather calm your mind and let the dream itself find its way through speech. And while you start translating it, you'd be amazed to see that you can interpret the dream and make proper associations between the dream content and your life events.

Chapter 5

Studies in Hysteria - Emotional Suppression and Its Dangerous Consequences

Are you a kind of person who likes to keep all your emotions hidden from the people around you! Do you prefer restraining your feelings a little too much! In that case, you must know that too much emotional suppression can have catastrophic impact over your body. The results are blindness, paralysis, numbness and other major neurological problems. This kind of physiological condition caused by emotional suppression is known as Conversion Disorder and the older term for it was "Hysteria". Now the word hysteria is commonly used to describe unmanageable emotional excess.

We get to know the essentials of hysteria from the joint work of Sigmund Freud and Josef Breuer, known as "Studien uber Hysterie" (Studies in Hysteria), first published in 1895. The book consists an introductory paper and five individual studies of hysterics (Breuer's famous case of Anna O. and four more by Freud). Before Freud and Breuer, it was the French neurologist, Jean-Martin Charcot who brought the attention of the world towards hypnosis and hysteria. He initially believed that hysteria was a neurological disorder for which patients were pre-disposed by hereditary features of their nervous system, but near the end of his life concluded that hysteria was a psychological disease. But the condition of Conversion Disorder or Hysteria dates back to ancient Greece.

Some doctors falsely believe that this disorder is not a real condition as it doesn't have any underlying biological cause, so they abruptly tell their patients that the problems do not exist in real, it's all in the patient's head. People with Conversion Disorder are not faking (malingering) their symptoms, although there is a syndrome in which a person feigns disease, illness or psychological trauma to draw attention and sympathy; it is known as Munchausen Syndrome.

Conversion disorder (functional neurological symptom disorder) is classified as one of the somatic symptom and related disorders in the Diagnostic and Statistical Manual of Mental Disorders of the American Psychiatric Association, Fifth Edition, (DSM-5). This can happen to anybody. The harder you try to keep your feelings inside, which you think as inappropriate to express, the more vulnerable you become to Conversion Disorder symptoms. For example; in an argument you get so intensely mad at someone that you want to hit that person, but as you think it might be totally inappropriate, you restrain yourself and don't let the anger out. The result of such extreme suppression can be catastrophic.

In similar events, many people have lost their ability to speak or have even become paralyzed. Does that mean when you feel like hitting someone, you should definitely do that! Not at all… there is another way to deal with this kind of situation. You just need to talk your feelings out to a close friend. How you feel! How much rage is bursting inside you! How much you want to pull off the person's head and play basketball with it! Spit it all out.

In most cases Hysteria symptoms occur because of any psychological conflict. The emotional battle that you feel within, the conflict between guilt and evil, rage and peace can turn the world of physical body upside down. Although Freud and Breuer were not first ones to work on Hysteria, their work made a real impact over the scientific community.

While Breuer, with his intelligent and amorous patient Anna O., had unwittingly laid the groundwork for psychoanalysis, it was Freud who drew the consequences from Breuer's case. The psychological conflict can have various factors underlying.

On this matter Freud had a huge misperception. He wanted a grand unifying theory for all hysteria symptoms. That's why he always was obsessed with his theory of sexual conflict underlying hysteria. Breuer on the other hand kept on finding different factors such as extreme emotional suppression, varieties of trauma etc.

This can happen to anybody, but documented literature shows more number of women suffering from Conversion Disorder symptoms than men. But why exactly, women are more vulnerable to such symptoms! All psychological states are the product of various hormonal interaction within the body. And a female body goes through way more hormonal changes in her lifetime than a male body. So it is easier for a psychological conflict to influence the hormonal interaction within a woman's body than a man's.

Hippocrates (5th century BC) was the first to use the term hysteria. Indeed he also believed that the cause of this disease lies in the movement of the uterus ("hysteron"). The Greek physician provides a good description of hysteria, which is clearly distinguished from epilepsy. He emphasizes the difference between the compulsive movements of epilepsy, caused by a disorder of the brain, and those of hysteria due to the abnormal movements of the uterus in the body. Then, he resumes the idea of a restless and migratory uterus and identifies the cause of the indisposition as poisonous stagnant humors which, due to an inadequate sexual life, have never been expelled. He asserts that a woman's body is physiologically cold and wet and hence prone to putrefaction of the humors (as opposed to the dry and warm male body). For this reason, the uterus is prone to get sick, especially if it is deprived of the benefits arising from sex and procreation, which, widening a woman's canals, promote the cleansing of the body. And he goes further; especially in virgins, widows, single, or sterile women, this "bad" uterus, since it is not satisfied not only produces toxic fumes but also tends to wander around the body, causing various kinds of disorders such as anxiety, sense of suffocation, tremors, sometimes even convulsions and paralysis. For this reason, he suggests that even widows and unmarried women should get married and live a satisfactory sexual life within the bounds of marriage. However, when the disease is recognized, affected women are advised not only to partake in sexual activity, but also to cure themselves with acrid or fragrant fumigation of the face and genitals, to push the uterus back to its natural place inside the body.

Even Aristotle and Plato seemed to have perceived the symptoms of Hysteria as a result of a sad uterus. Somehow all these geniuses of history had the idea that hysteria was related to the lack of sexual pleasure. Plato, in Timaeus, argues that the uterus is sad and unfortunate when it does not join with the male and does not give rise to a new birth.

But don't worry, we are not going to mess with anybody's uterus. That might be considered as sexual harassment. So, my dear Hippocrates, Plato and Aristotle... thanks... but no thanks.

Modern neuroscience shows us that it's not really about whether the uterus is sad, rather it's about whether you are sad, and yet not sharing your sorrow with anyone. A sad uterus is just a symbolism for the dissatisfaction in the sexual life of a woman. This implies that sexual suppression may come out as physical problems, if you cross the line of repression. But again, sexual repression is just one among many factors in Conversion disorder symptoms. There are tons of other factors that can cause those symptoms. For example a sudden mental trauma can leave a person totally blind or mute.

French neuropsychiatrist Pierre Janet (1859-1947), with the sponsorship of J. M. Charcot, opened a laboratory in Paris' Salpetriere. He convinced doctors that hypnosis based on suggestion and dissociation was a very powerful model for investigation and therapy. He wrote that hysteria is *"the result of the very idea the patient has of his accident"*. The patient's own idea of pathology is translated into a physical disability. Janet studied five hysteria symptoms: anaesthesia, amnesia, abulia, motor control diseases and modification of character. She suggested that the reason of hysteria is in the *"idee fixe"*, that is the subconscious. Janet's studies are very important for the early theories of Freud, Breuer and Carl Jung.

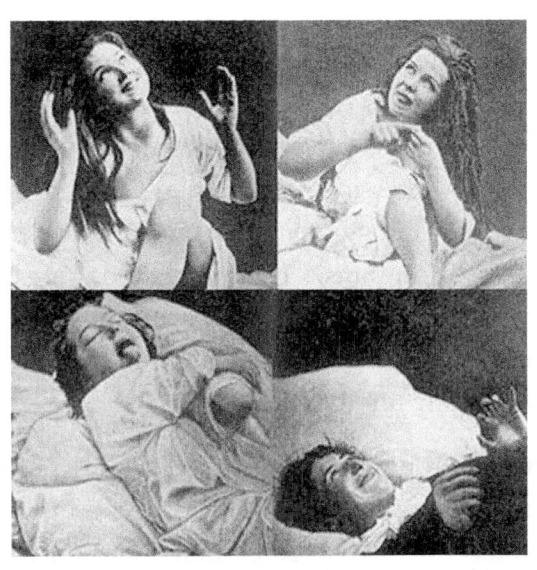

Figure 5.1 A Hysteria Patient Under The Effects of Hypnosis

The most interesting case in hysteria's history was Breuer's patient Bertha Pappenheim, alias Anna

O. According to Breuer, Anna *"had hitherto been consistently healthy and had shown no signs of neurosis during her period of growth. She was markedly intelligent, with an astonishingly quick grasp of things and penetrating intuition. She had great poetic gifts, which were under the control of a sharp and critical common sense"*.

Figure 5.2 Breuer's Patient Anna O.

Breuer reported, Anna fell prey, during her father's final illness and in the months after his death, to the most appalling symptoms of hysterical paralysis and anaesthesia in three out of her four limbs, together with a succession of other distressing psychiatric symptoms.

At different times these included weakness, inability to turn her head, diplopia, a nervous cough, loss of appetite, hallucinations, agitation, mood swings, abusive and destructive behavior, amnesia, somnolence, tunnel vision and partial aphasia

Breuer further recorded: *"She no longer conjugated verbs and eventually she used only infinitives, for the most part incorrectly formed from weak past participles"*. Among her symptoms, she was at one time unable to speak in her native German, but could still read both French and Italian, translating them aloud into English. On one occasion she was for several weeks unable to drink in spite of a tormenting thirst. Often she experienced alterations in her personality accompanied by confusion and delirium, this state was called "absence".

Breuer observed that, while the patient was in her state of absence, she was in the habit of muttering a few words to herself which seemed as though they arose from some train of thought that was occupying her mind. During part of her illness, she was unable to recognize or accept food from anyone except her physician, who spent almost a total of a thousand hours with her between April 1881 and June 1882. She was able to satisfy herself of his identity only by holding his hands.

Notice that all her symptoms were not the product of any organic cause, rather they were totally psychological. Breuer figured out that as Anna started to talk out all about her fairy tales, previously repressed feelings and thoughts, her hysteria began to disappear. She coined the term "talking cure" for this kind of talking treatment. Sometimes jokingly she used to call this "chimney sweeping". Later these simple words of Anna O. got their sophisticated forms as "psychotherapy" or "counselling".

Each evening Breuer would return and Anna would recount, with vivid emotion, the exact events from precisely one year ago. In the final stage, Anna began to add to these accounts a description of the various occurrences that had evidently triggered each of her hysterical symptoms during the previous year. As she did so, the relevant symptom itself would disappear. For example, on recalling her disgust at seeing a dog drink from a lady companion's glass of water a year before, she was suddenly able to drink once more. She recovered from her symptoms over time, and later in life she became a distinguished social worker and a noted campaigner for women's rights.

But let's not use the term "Hysteria", as it is not the modern day term for the disorder. So, we are going to use the term Conversion Disorder while exploring the neurobiology behind it. Conversion disorder is a specific form of somatization in which the patient presents with symptoms and signs that are confined to the voluntary central nervous system.

Now, what is "Somatization"?

Somatization is the psychological mechanism whereby psychological distress is expressed in the form of physical symptoms. The psychological distress in somatization is most commonly caused by any kind of psychological conflict that threatens mental stability. Conversion disorder occurs when the somatic presentation involves any aspect of the central nervous system over which voluntary control is exercised. However no specific neurological pathway is discovered, that acts as a bridge between the psychological conflict and physical symptoms like Anaesthesia, Paralysis, Ataxia, Tremor, Partial Seizures, Deafness, Aphonia, Globus hystericus.

But still, several studies have given us amazing leads on the neurobiology of Conversion Disorder. As there is emotion involved in the Conversion symptoms, it is not farfetched to say that the original emotion center (presently we consider only the amygdala as the emotion center) of the brain, the limbic system is deeply involved.

Preliminary data from neuroimaging studies provide us evidence on possible networks engaging the limbic system and motor regions that may be involved in conversion symptoms like paralysis. Studies demonstrate the engagement of regions in the limbic-motor interface to attempted or imagined movement (ventromedial prefrontal cortex) and non-noxious brush stimuli (caudate/putamen) in conversion paralysis. These regions have been suggested as potential nodal points for emotional stimuli to influence motor function.

As a part of the limbic system, amygdala handles most of the emotional processes. There are several mechanisms by which amygdala activity may modulate motor behaviors. In a study by Valerie Voon et. al., they demonstrated aberrant limbic-motor interactions in patients with conversion disorder that may underlie the influence of affect or arousal on motor function.

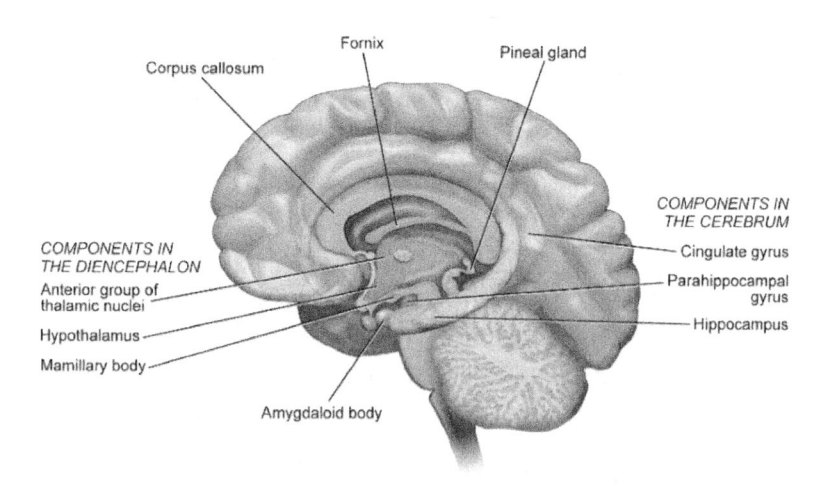

Figure 5.3 The Limbic System

Patients with motor conversion disorder had greater functional connectivity from the right amygdala to the right supplementary motor area, a region involved in motor initiation. Thus any imbalance in the emotional state can have potential impairing effects on the body. This explains how a trauma from a sudden death of the spouse can leave a person paralyzed or even catatonic.

So, the bottom-line is:

Do not suppress your emotions too much.

You won't even notice when you just cross the line and step into the catastrophic domain of Conversion Disorder. It's better not to hold your feelings inside too much and express them to a dear one freely, than to pay thousands of dollars to a psychiatrist for the same outburst of emotions later. Emotions are a bonding mechanism for humans. So, use 'em, abuse 'em and utilize 'em.

Chapter 6

Earth-Brain Bondage - Psychological & Physiological Changes during Fluctuations in Geophysical Parameters

Mother earth has put all her excellence in molding the human brain throughout the millions of years long evolutionary period. Just like, a mother's genetic traits are forwarded to her children, our beloved mother earth with the use of Darwinian concept of adaptation and naturally selective pressure, has gifted us an amazing and marvelous neural network of the brain which is capable of contemplating the universe, the limitless cosmos, the beauty of mother nature, and even capable of contemplating itself contemplating. The beautiful human brain is even able to feel what our planet is feeling.

Have you ever thought, why and how patients suffering from schizophrenia and paranoia happen to have more psychotic breaks at specific periods of the lunar cycle! This happens due to the disturbance in the geomagnetic field of planet earth caused by lunar gravity. Likewise a wide variety of cerebral, biological responses and medical complications has been associated with geophysical events. These effects include alterations in occurrences of migraine headaches, strokes, glaucoma pains, joint swelling (objectively measured), blood clotting, skin conductivity, tissue permeability, embolism risks, thyroid activity, heart attacks etc. In this chapter I'll describe how deeply our beloved mother earth is ingrained within our brain or to talk scientifically, I'll explain the interconnectedness between the earth's geomagnetic activity and the cerebral activity. I'm about to show you the fantastic craftsmanship of planet earth while molding the 3 lbs. lump of jelly.

The human brain is perhaps the most intricate and advanced creation (not from a creationism perspective, but from a completely evolutionary) of Mother Nature on earth. At present times, the mesmerizing exploration of mysteries within the human brain is still in infancy. We have only started to take baby steps.

If we go deeper into the spectacular network of the human brain, all we shall experience is a feeling of awe. At every level of the mesmerizing neural circuitry we shall discover an undeniable signature of Mother Nature or more specifically planet earth. Humans have not just evolved on this planet, indeed, the planet molded the brain as per its own specific criteria and even sowed the seeds of its own characteristics deep within the neural network.

During a long period of 15 million years, since the time when Hominids (Great Apes) diverged from the Gibbon family, evolution has slowly made us everything we are. As a result of that prolonged process of biological evolution we see mysterious neurological conditions occurring during global or solar events, which only make sense when we take various geophysical and cosmological factors into account.

Human brain is the most complicated mystery of our planet. The womb of mother earth is filled with mysterious mechanisms. These mechanisms make the creatures of this planet uniquely special in one way or another. Every living organism on earth is always connected to the geophysical state of the planet. We see the beauty of this correlation everywhere around us, most apparently in animals. For example, an accurate navigation system has been gifted to various earthly creatures by our planet, which is called "Magnetoreception". It works in pace with the earth's geomagnetic field and helps the migratory birds to find their way back home.

To explain a little more we can say that the avian magnetic compass is a complex biological mechanism with many surprising properties. The basis for the magnetic sensing is located in the eye of the bird, and furthermore, it is light-dependent, i.e., a bird can only sense the magnetic field if certain wavelengths of light are available. Specifically, many studies have shown that birds can only orient if blue light is present.

Likewise, animal behavior prior to earthquakes is quite captivating to human imagination. All animals instinctively respond to escape from predators and to preserve their lives. A wide variety of vertebrates already expresses "early warning" behaviors right before an earthquake. So it's plausible that a seismic-escape response could have evolved from this already-existing genetic predisposal. An instinctive response following a Primary wave seconds before a larger Secondary wave is not a huge leap, so to speak, but what about other precursors that may occur days or weeks before an earthquake! In fact there are precursors to a significant earthquake that we have yet to learn about, such as ground tilting, groundwater changes, electrical or magnetic field variations. Indeed it's possible that some animals could sense these signals and connect the cognitive phenomenon with an impending earthquake.

Now let's focus on the main mettle of this chapter. What kind of earthly signature the human brain has, without even being aware of it. Homo sapiens brain has been immersed in the geomagnetic field of planet earth from the very beginning of human evolution. As a result every single neural firing in the synapses has been interacting with the surrounding geomagnetic field at all times. If we go back in time we'll find out that, a major adaptive advantage of human evolution was an amazing increase in brain size. Fossil evidence allows us to trace the development of the brain as it increased threefold over the last three million years. Throughout human evolution, the brain has continued to expand.

The earliest documented members of the genus Homo are Homo habilis, which evolved around 2.4 million years ago. This was the earliest species for which there is positive evidence of use of stone tools. The brains of these early hominins were about the same size as that of a chimpanzee, although it has been suggested that this was the time in which the human SRGAP2 gene doubled, producing a more rapid wiring of the frontal cortex, that is responsible for analytical thinking.

During the next million years a process of encephalization began, and with the arrival of Homo erectus in the fossil record, cranial capacity had doubled from 550 cc to around 1100 cc. This increase in human brain size is equivalent to every generation having an additional 125,000 neurons more than their parents.

Homo erectus were the first species in the history of planet earth that learnt to tame and create fire. They also excelled in making complex tools. Throughout this entire human evolution period planet earth had long enough time to mold the human brain and make it most suitable to survive and further evolve on this planet.

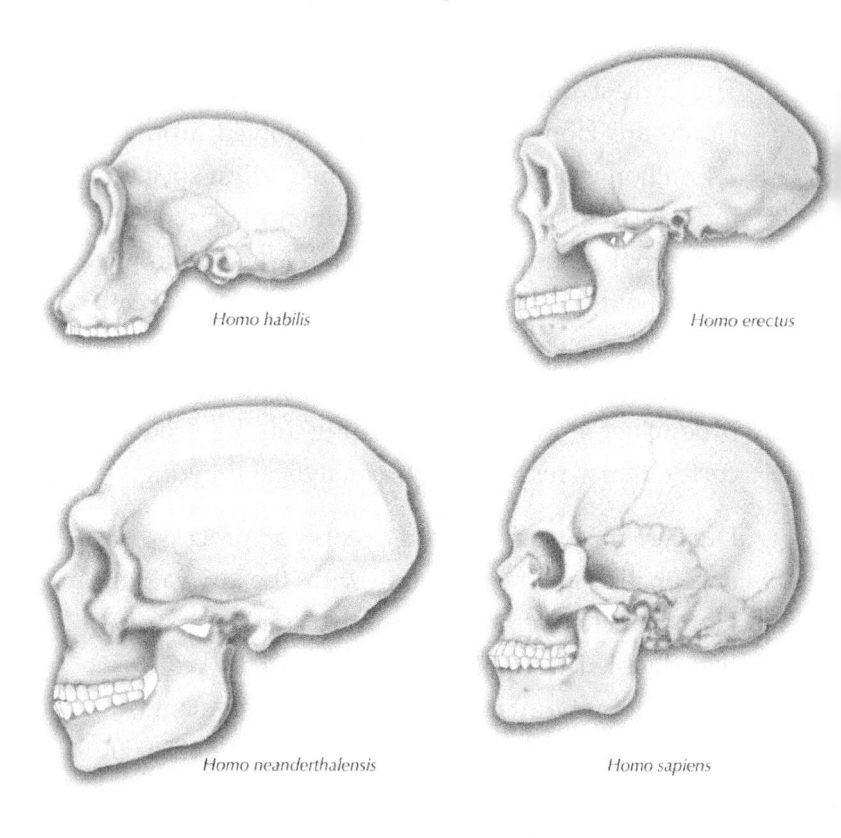

Figure 6.1 Series of hominid skulls with different brain sizes (Homo habilis 550 cc, Homo erectus 1100 cc, Homo neanderthalensis 1500 cc, Homo sapiens 1300 cc)

EEG pattern of the brain waves are the most eloquent evidence of planet earth's geomagnetic touch over cerebral activity. Lewis B. Hainsworth of Western Australia seems to be the first researcher to recognize the relationship of brain wave frequencies to the naturally circulating rhythmic signals of the planet, known as Schumann's Resonance (SR).

In 1952, a German physicist, Dr W.O. Schumann, suggested that the space between the surface of the earth and the ionosphere acts as a resonant cavity, somewhat like the chamber in a musical instrument. Energy for the SR is provided by lightning. Lightning pumps energy into the earth-ionosphere cavity and causes it to resonate at frequencies in the ELF range.

Schumann calculated the SR frequencies and fixed the most predominant standing wave at about 7.83 Hz. The frequency values of the SR signals are determined by the effective dimensions of the cavity between the Earth and ionosphere. Thus, any kind of event that changes these dimensions will change the resonant frequencies. Such events could be ionospheric storms, and could even result from a man-made ionospheric disturbance. These ionospheric disturbances whether they are man-made or caused by a solar-storm (Coronal Mass Ejection), produce significant geomagnetic storms and anomalies. These geomagnetic conditions change diurnally, seasonally and with variations in solar activity, which, in turn varies with the 11 year sunspot cycle and also with the 27-29 day lunar cycle, mainly during sunspot minimum periods.

Hence, these changes in the atmospheric conditions sometimes affect the biological systems. For instance, lunar cycle creates geomagnetic disturbance, which in turn proves to be the precursor and stimulant to rheumatoid arthritic pain in some aged people. A number of biologists have concluded that the frequency overlap of SR and biological fields is not accidental, but is the culmination of a close interplay between geomagnetic and biomagnetic fields over evolutionary time. Hence researchers have examined interactions between external fields and biological rhythms. Organisms are capable of sensing the intensity, polarity and direction of the geomagnetic field. A variety of behavioral disturbances in the human population is statistically related to disturbances in the geomagnetic field :

Friedman et al (1965) documented a relationship between increased geomagnetic activity and the rate of admission of patients to 35 psychiatric facilities.

Venkatraman (1976) and Rajaram & Mitra (1981) reported an association between changes in the geomagnetic field due to magnetic storms and frequency of seizures in epileptic patients.

Geophysical features determine the frequency spectrum of human brain wave rhythm. Moreover, the frequencies of naturally occurring electromagnetic signals, circulating in the electrically resonant cavity bounded by the Earth and the ionosphere, have governed or determined the evolution or development of the frequencies of operation of the principal human brain wave signals. Which means, the geophysical parameters nourished and developed the human brain, and in fact the entire anatomy, to live in the tuned system of planet earth by means of the naturally occurring Schumann's resonances.

This way, the magnetic and electric field strength of the human species have approached the values same as Schumann's Resonances. Therefore, based on the facts, I deduce that: *human brain is an intricate, miniaturized and vivacious projection of planet earth itself.* Any disturbance in the geophysical parameters is meant to be sensed by the human brain. Or to put it more simply, if anything happens to our beloved planet, no matter how subtle it is, our evolutionary instincts make us sense it.

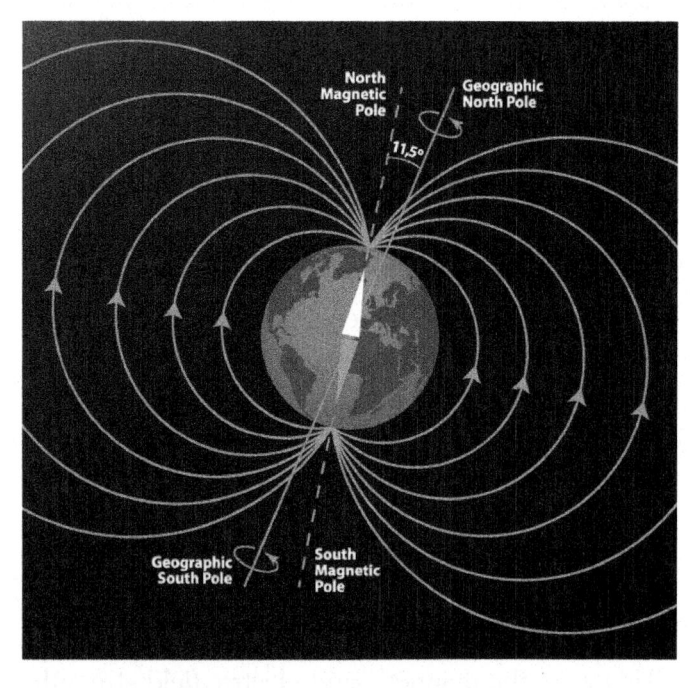

Figure 6.2 Earth's Geomagnetic Field

Every species is special and gifted in its own way. The earth-ionosphere cavity acts as the womb of mother earth, where she nourishes her children and sees them grow intellectually millennium after millennium. The only difference between mother earth and a human mother is that, after the umbilical cord is cut, the direct physical attachment is discontinued between the mother and her child, while on the other hand, human species is still too immature to break the earth-brain bondage. Studies have shown that subjects living in isolation from geomagnetic rhythms over long period of time developed increasing irregularities and chaotic physiological symptoms, which were dramatically restored after the introduction of a very weak SR electromagnetic field. Early astronauts suffered behavioral disturbances until SR generators were installed in the spacecrafts.

A mother usually does not show more affection to one child and less to the other. But, on the contrary mother earth definitely loves humans more than any other species. The most significant evidence for this is the emergence of an inexplicable intellect in humans. And the geomagnetic field allows the humans to nourish this intellect, which is a product of millions of years of evolution.

The evolution of the physical, chemical and biological phenomena that have aggregated to produce complex organisms and the electromagnetic processes that emerged within their spatial boundaries have occurred while immersed in the geomagnetic field and the intensities and frequencies generated as Schumann resonances within the earth-ionosphere cavity. The remarkable convergences between the temporal patterns and the intensities of the magnetic and electric field components for Schumann resonances found within both global human quantitative electroencephalographic activity and earth-ionosphere activity quite simply shows the significant existence of the earth-brain bondage. Just like mirror neurons are the basis of the human capacity to empathize with another person, the brain's own electromagnetic characteristics enable the humans to empathize with our planet earth.

Chapter 7

GOD is A Figment of Your Imagination – Experiencing GOD in The Brain

God is the most undeniably important, perpetual delusion of human evolution. The fact that the concept of God has survived so far in human minds while evolving with the human brain, implies its significance eloquently. Primitive men worshiped the mountains, the sun, thunder and as their cerebral neural network evolved, they started to worship more obvious humanlike forms of Gods, along with a few religious sectors of the society believing in the Supreme Energy source which has no form.

But this chapter is not at all about religion. It is neither on the argument about whether God exists or not. Rather, in this chapter I'll show the scientific basis of the concept of God, how it is hardwired within the human brain and how exactly it has been and still is necessary for most of the earthlings!!!

Well; to understand this extraordinarily long process we need to go back in time and look behind the curtain, how the Darwinian evolution paved the way for this mysterious concept, and more importantly how exactly some of the most important God-like experiences occurred in the human brain throughout mankind history! Experiences like Buddha's Nirvana, Moses's Ten Commandments, Muhammad's Revelation, Indian Sages' Vedas & Upanishad and many others have fueled the philosophical needs of mankind for ages.

Figure 7.1 Moses and The Ten Commandments

God is nothing but an essential figment of your imagination. All real and imaginary experiences are generated by the brain; experiences of all Sentient Beings, including God, are generated by cerebral activity. Mystics throughout history have claimed to experience visions and trans-like states, which they say come directly from God. But do they really come from God! In this chapter, I'll take you on a journey to explore the mysterious cerebral functioning behind religious/spiritual/mystical experiences (RSMEs).

Recorded history of neurological evidence shows that RSMEs are often evoked by transient, electrical microseizures within the temporal lobes of human brain. Normally when we think of epileptic seizures, we think of someone convulsing and losing consciousness. But that's just the most known one type of seizure called grand mal seizure.

There's a whole other category of seizures, known as focal or partial seizures, that can cause a kaleidoscope of symptoms, such as the sense of oneness, complex hallucinations and feelings of fear, depression, and euphoria. Often, these seizures don't involve any convulsions at all. In some epileptics, a seizure can even invoke the presence of God.

Figure 7.2 Fyodor Mikhailovich Dostoyevsky

The Russian Novelist, Fyodor Dostoyevsky, whose writings are among the world's greatest literature, had a rare form of temporal lobe epilepsy termed "Ecstatic Epilepsy". Dostoyevsky kept records of 102 epileptic seizures during his last two decades, which mainly occurred at night. Seizures which occurred in the daytime were often preceded by an ecstatic aura, which has led neurologists to theorize that he had temporal lobe epilepsy. Based on his experiences, he created characters with epilepsy in his four novels The Possessed, The Brothers Karamazov, The Insulted and Injured, and The Idiot. The Idiot is an example of how art can contribute to scientific observation.

Dostoyevsky lets us see into the mind and emotion of the person with epilepsy through his character Prince Myshkin:

"He [Myshkin] remembered that during his epileptic fits, or rather immediately preceding them, he had always experienced a moment or two when his whole heart and mind, and body seemed to wake up to vigour and light, when he became filled with joy and hope, and all his anxieties seemed to be swept away forever; these moments were but presentiments, as it were, of the one final second (it was never more than a second) in which the fit came upon him".

Dostoyevsky even recorded his own seizure experiences :

"For several instants I experience a happiness that is impossible in an ordinary state, and of which other people have no conception. I feel full harmony in myself and in the whole world, and the feeling is so strong and sweet that for a few seconds of such bliss one could give up ten years of life, perhaps all of life.

I felt that heaven descended to earth and swallowed me. I really attained god and was imbued with him. All of you healthy people don't even suspect what happiness is, that happiness that we epileptics experience for a second before an attack."

Likewise, historical data indicates that Joan of Arc, The Maid of Orleans from France, who is well known as the messenger of God was also an epileptic. She suffered from tuberculosis with a temporal lobe tuberculoma and tuberculous pericarditis. It is possible to explain Joan of Arc's experiences and behavior in terms of a widespread chronic tuberculous infection which became inactive in some organs' and calcified in others.

Figure 7.3 Joan of Arc Being Burnt at the Stake

Calcification of the tuberculous mesenteric glands and chronic tuberculous pericarditis could account for the heart and parts of the intestines being intact after she was burnt at the stake. A tuberculoma (lesion) in the temporal lobe of her brain could account for her complex visual and auditory hallucinations of Archangel Michael, Saint Margaret and Saint Catherine.

Until now there have been countless events of such hallucinatory God-encounters or experiences among the human population of the world. It was in the early 19th Century, Paris when it got first recognized by the scientific community that epilepsy is the root of all religious/spiritual/mystical experiences (RSMEs).

Every now and then we neuroscientists come across new speculations of mystical experiences. A specific conversion experience after an epileptic fit was reported in 1872. The patient believed that he was in Heaven. He would appear to have been depersonalized, as it took three days for his body to be reunited with his soul according to the patient. He expressed that he was now a new man, and had never before known what true peace was.

Most of the Temporal Lobe Epilepsy patients express the feeling of inexplicable bliss during the seizure. Many hear the voice of God. Everything starts to seem interconnected to them, and they finally understand the meaning of the universe. Many of them declare that God has given them different missions to complete, like building a temple or church or even reforming the whole world. They endorse strong personal beliefs in either specific cultural deities such as Allah, Brahma, Vishnu, Shiva, Christ, Virgin Mary or more exotic god surrogates such as "The Supreme Universe". But all these experiences are happening inside the brain, nowhere outside. It is just our mind playing tricks on us.

Just like a lesion in the temporal lobe, naturally occurring electromagnetic fields can cause RSMEs as well by impacting over the cerebral activity in the temporal lobe. For example, powerful meteor showers were occurring when Joseph Smith, founder of the Church of Latter Day Saints, was visited by the angel Moroni, and when Charles Taze Russell formed the Jehovah's Witnesses. The experience of God and all spirits can even be induced by artificial electromagnetic stimulation around the head.

Many neuroscientists have tried to find out the God-Spot in the brain that is responsible for RSMEs. But what we have found is that it is not just one, rather several major regions in the brain that are responsible for all kinds of God experiences.

Andrew Newberg, a neuroscientist at the University of Pennsylvania used single photon emission computed tomography (SPECT) to take pictures of the brain during religious activity. SPECT provides a picture of blood flow in the brain at a given moment, so more blood flow indicates more activity. One of Newberg's studies examined the brains of Tibetan Buddhist monks as they meditated. The monks indicated to Newberg that they were beginning to enter a meditative state by pulling on a piece of string. At that moment, Newberg injected radioactive dye via an intravenous line and imaged the brain. Newberg found increased activity in the frontal lobe, which deals with concentration; the monks obviously were concentrating on the activity.

But Newberg also found an immense decrease of activity in the parietal lobe. The parietal lobe, among other things, orients a person in a three-dimensional space. This lobe helps you look around to determine that you're 15 feet away from the window, 6 feet away from the door and so on. Newberg hypothesizes that the decreased activity in the brains of the meditating monks indicates that they lose their ability to differentiate where they end and something else begins. In other words, they become one with the universe, a state often described in a moment of transcendence or Samadhi.

Newberg found similar brain activity in the brains of praying nuns. So, it does not to whom or what that religious activity is directed toward. Though the nuns were praying to God, rather than meditating like the monks, they showed increased activity in the frontal lobe as they began focusing their minds. There was also a decrease of activity in the parietal lobe, seemingly indicating that the nuns lost their sense of self in relation to the real world and were able to achieve communion with God.

The brain has two hemispheres. The left hemisphere is more of a logical and practical person, while the right hemisphere is the creative genius that creates and senses the supernatural. All your sense of self and the sense of the other are derived from the subtle but complex structural and neuroelectrical differences between the left and right hemispheres of the brain.

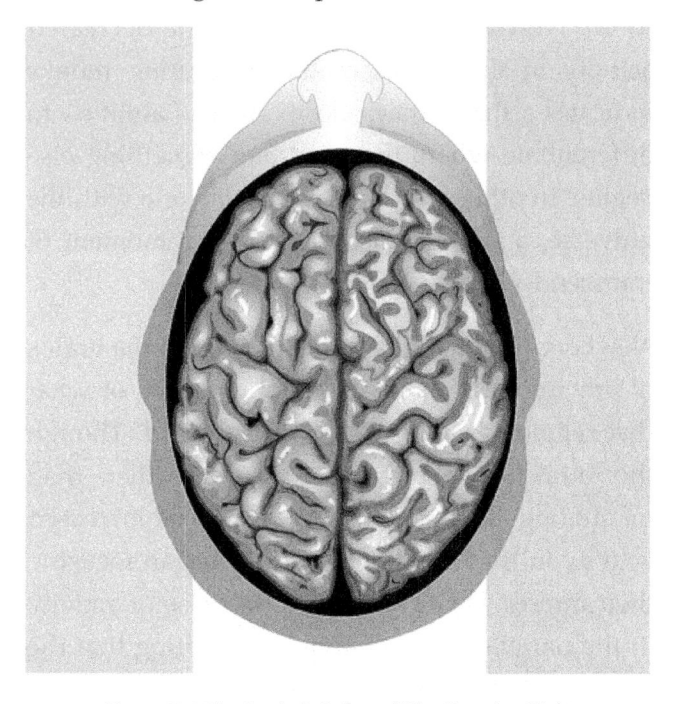

Figure 7.4 The Analytic Left and The Creative Right Hemisphere of The Brain

Whereas traditional left, more linguistic, hemispheric processes are strongly coupled to the sense of self, the transient intrusion of the right hemispheric equivalents are associated with the sensed presence of the supernatural, guiding angel or the supreme. So, basically you have two faces of your mind, but only one of them remains dominant in every person. Literally, half your brain is theist and the other half is atheist. So, what really happens when you die! Does half your brain go to heaven and the other half go to hell!!!

But jokes apart; human brain is hardwired to find meaning in everything it observes. That's why when you see an absurd painting, you can find your own meaning in it. A hundred people looking at a Van Gough would see the painting in a hundred different ways. It is inherent in human brain to make sense even out of nonsense. It is an evolutionary mechanism that developed through millions of years to keep the mind calm even in confusing times. When the brain lacks information on the scientific explanation for something, it tends to fill the gaps with supernatural explanation. It's neither stupid, nor foolish, rather it's really human to believe in the supernatural. Just like atomic energy has the potential to power a big city as well as to destroy one, God-experience has the potential to make better and confident human beings as well as to kill thousands of people in events like 9/11.

But does this mean that all the religious founders and the spiritual giants of human history were epileptic! The answer would be a straight no. Most of them had their own God-experiences through the natural means of prayer, devotion and meditation. On the other hand, temporal lobe microseizures triggered by brain lesion or planet earth's geomagnetic disturbances can evoke similar experiences as the religious founders.

However, whatever the cause may be, in all cases the deep spiritual experience takes place right within the complicated mesh of nerve cells of the human brain, without the intervention of any Almighty Being. And the content of the experience is formed by the brain based upon the person's own emotions, beliefs, intuitions, urges and conjectures. Thus the person's own distinct characteristics are imposed on the manifestation of the experience. Moreover, various neurochemical changes during the deep spiritual experience, leaves the person radically transformed. Often, the person comes out of the experience as a completely changed human being with certain unbreakable beliefs and intuitions about the universe. And all of this happens right within the neurons.

Now, one might wonder, what about the rest of the general human population who tend to be spiritual or religious by nature? Do all of them experience some sort deep spiritual encounter?

In majority of the general religious/spiritual population, what we see is more of a basic sense of religiosity/spirituality, than a deep spiritual conviction like the religious founders or the temporal lobe epileptics. Indeed they are religious/spiritual in their hearts. However, that is also not by choice, but driven by the evolutionary instincts. Darwinian natural selection has embedded the characteristic of religious belief in the brain circuits to survive the hardship of daily life.

Also, as humans spend the beginning of their lives under the guidance and care of parents, it is ingrained in their involuntary reflex to rely on a parental guidance all through the lifetime, either as a father figure like Jesus Christ or as a mother like Goddess Kali, or as a more abstract higher power such as Paramatman or Allah. Sometimes it's just good to know that someone or something is looking after you, guiding you in the right path and giving you the strength to survive the bad times.

Bibliography

Aserinsky E & Kleitman N. "Regularly occurring periods of eye motility, and concomitant phenomena, during sleep". Science 118, 1953.

Allman et. al. (2001). "The anterior cingulate cortex..." Ann N Y Acad Sci. 935

Apps M., Balsters J, Ramnani N, (2009). "Anterior Cingulate Cortex: Monitoring The Outcomes Of Others' Decisions", Royal Holloway University of London, London, United Kingdom, http://www.medicalnewstoday.com/articles/153472.php

Ardila A, Gomez J. Paroxysmal "feeling of somebody being nearby". Epilepsia 1988;

Babayev ES, Allahverdiyeva AA. Effects of geomagnetic activity variations on the physiological and psychological state of functionally healthy humans: some results of Azerbaijani studies. Adv Space Res 2007.

Burchett, Scott A. and Phillip T. Hicks. 2006. "The mysterious trace amines: Protean neuromodulators of synaptic transmission in mammalian brain." Progress in Neurobiology.

Balleine BW, Delgado MR, Hikosaka O (2007a). "The role of the dorsal striatum in reward and decision-making". J Neurosci 27

Barrett, D. (2001) The Committee of Sleep: How Artists, Scientists, and Athletes Use their Dreams for Creative Problem Solving—and How You Can Too. NY: Crown Books/Random House/hardback

Barrett, D. (2007) "An Evolutionary Theory of Dreams and Problem-Solving" in Barrett, D.L. & McNamara, P. (Eds.) The New Science of Dreaming, Volume III: Cultural and Theoretical Perspectives on Dreaming, NY, NY: Praeger/Greenwood, 2007.

Bechara, A. et al (1994) "Insensitivity to future consequences following damage to human prefrontal cortex". Cognition 50

Blackmore, Susan (2004). Consciousness an introduction. New York, NY: Oxford University Press

Braun, AR (1997) "Regional cerebral blood flow throughout the sleep-wake cycle. An H2(15)O PET study". Oxford Journals, Brain; (1997) 120 (7).doi: 10.1093/brain/120.7.1173

Bottini G, Corcoran R, Sterzi R, Paulesu E, Schenone P, Scarpa P, et al. (1994) "The role of the right hemisphere in the interpretation of figurative aspects of language". Brain 117

Botvinick M, et al (1999) "Conflict monitoring versus selection-for-action in ACC" Nature 402 (6758): 179–81

Bush G, Luu P, Posner MI. (2000). "Cognitive and emotional influences in anterior cingulate cortex". Trends Cogn Sci. 2000 Jun; 4(6)

Blanke, O. and Arzy, S. "The Out-of-Body Experience: Disturbed Self-Processing at the Temporo-Parietal Junction" THE NEUROSCIENTIST 2005.

Blanke O, Ortigue S, Landis T, Seeck M. 2002. Stimulating illusory own-body perceptions. Nature.

Bonda E, Petrides M, Frey S, Evans A. 1995. Neural correlates of mental transformations of the body-in-space. Proc Natl Acad Sci U S A.

Brandt T. 2000. Central vestibular disorders. In: Vertigo: its multisensory syndromes. London: Springer.

Brugger P, Agosti R, Regard M, Wieser HG, Landis T. 1994. Heautoscopy, epilepsy, and suicide. J Neurol Neurosurg Psychiatr.

Brugger P, Regard M, Landis T. 1997. Illusory reduplication of one's own body: phenomenology and classification of autoscopic phenomena. Cogn Neuropsychiatr.

Breur, J. and Freud, S. (1893-1895) Studies on Hysteria.

Carta, M.G. "Women And Hysteria In The History Of Mental Health" Clinical Practice & Epidemiology in Mental Health, 2012

Cartwright, R. (1993). "Functions of Dreams". Encyclopedia of Sleep and Dreaming.

Calvert GA, Campbell R, Brammer MJ. 2000. Evidence from functional magnetic resonance imaging of crossmodal binding in the human heteromodal cortex. Curr Biol.

Campbell A. The limbic system and emotion in relation to acupuncture. Acupuncture in Medicine.1999.

Cook, R. "Mirror neurons: From origin to function" Behavioral and Brain Sciences 2014.

Daly DD. 1958. Ictal affect. Am J Psychiatry.

Darwin, Charles. "On the origin of species by means of natural selection" (original edition, 1859).

Darwin, Charles. "The Descent of Man" (original edition, 1871).

Decety J, Sommerville JA. 2003. Shared representations between self and other: a social cognitive neuroscience view. Trends Cogn Sci.

Denning TR, Berrios GE. 1994. Autoscopic phenomena. Br J Psychiatry.

Devinsky O, Feldmann E, Burrowes K, Bromfield E. 1989. Autoscopic phenomena with seizures. Arch Neurol.

Downing PE, Jiang Y, Shuman M, Kanwisher N. 2001. A cortical area selective for visual processing of the human body. Science.

Dang-Vu et.al., (2007) "Neuroimaging of REM Sleep and Dreaming", In D. Barret & P. McNamara (Eds.), The New Science of Dreaming: Vol. 1. Bilogical Aspects. Westport Connecticut, Praeger.

Dastur H.M., Desai A.D., "A comparative study of brain tuberculomas and gliomas based upon 107 case records of each". Brain. 1965

Devinsky O, Lai G. Spirituality and religion in epilepsy. Epilepsy Behav 2008.

Dostoyevsky, The Possessed (original edition 1872)

Dostoyevsky, The Brothers Karamazov, (original edition 1880)

Dostoyevsky, The Insulted and Injured, (original edition 1861)

Dostoyevsky, The Idiot,. (original edition 1869)

Dotta BT, Persinger MA. "Doubling" of local photon emissions when two simultaneous, spatially separated, chemiluminescent reactions share the same magnetic field configurations. J Biophys Chem 2012

Dotta BT, Bucner CA, Lafrenie RM, Persinger MA. Photon emissions from human brain and cell culture exposed to distally rotating magnetic fields shared by separate light-stimulated brains and cells. Brain Res 2011

Esquirol, Étienne (1838). Baillière, Jean-Baptiste (and sons), ed. Des maladies mentales considérées sous les rapports médical, hygiénique et médico-légal, [Mental illness as considered in medical, hygienic, and medico-legal reports] Volume 1 and 2.

Esquirol, Étienne 1845. Mental maladies; a treatise on insanity (original French edition 1838).

Esch, T. and Stefano, G.B "The Neurobiology of Love" Neuroendocrinology Letters No.3 June Vol.26, 2005.

Esch T. [Health in stress: Change in the stress concept and its significance for prevention, health and life style]. Gesundheitswesen 2002.

Flaubert Gustave, Madame Bovary 1856

Feinstein, D. (1990) "The Dream as a Window to Your Evolving Mythology," in S. Krippner, Dreamtime & Dreamwork, Jeremy P. Tarcher Inc. Los Angeles, CA

Farrell MJ, Robertson IH. 2000. The automatic updating of egocentric spatial relationships and its impairment due to right posterior cortical lesions. Neuropsychologia.

Fasold O, von Bevern M, Kuhberg M, Ploner CJ, Vilringer A, Lempert T, Wenzel R. 2002. Human vestibular cortex as identified with caloric vestibular stimulation by functional magnetic resonance imaging. Neuroimage.

Fiss, (1986)., "An experimental self-psychology of dreaming" Journal Of Mind And Behavior 1986

Foulkes, D. (1982) Children's dreams: longitudinal studies, Wiley, 1982

Fosse MJ, Fosse R, Hobson JA, Stickgold RJ. (2003). "Dreaming and episodic memory: a functional dissociation?" J Cogn Neurosci. 2003 Jan 1

Freud, S. (1900). The Interpretation of Dreams

Freud, S. "Selected papers on hysteria and other psychoneuroses" Journal of Nervous and Mental Disease 1909.

Freud, S. "The Origin and Development of Psychoanalysis" (1910)

Freud, S. (1914) Psychopathology of everyday life

Goodwin GM, McCloskey DI, Matthews PBC. 1972. Proprioceptive illusions induced by muscle vibration: contribution by muscle spindles to perception? Science.

Green CE. 1968. Out-of-body experiences. London: Hamish Hamilton.

Grossman E, Donnelly M, Price R, Pickens D, Morgan V, Neighbor G, and others. 2000. Brain areas involved in perception of biological motion. J Cogn Neurosci.

Grüsser OJ, Landis T. 1991. The splitting of "I" and "me": heautoscopy and related phenomena. In: Visual agnosias and other disturbances of visual perception and cognition. Amsterdam: MacMillan.

Greenberg, R., Pearlman, C. (1975). "Rem Sleep and the Analytic Process", Psychoanal Q., 44

Greene, G. (2010) "The Power And Purpose Of Dreams", Psychology Today Published on February 15, 2010

Griffin, J. (1997). "The Origin of Dreams: How and why we evolved to dream". The Therapist 4

Gribbin John. (2013) "Erwin Schrodinger and The Quantum Revolution", Wiley

Goetz CG (August–September 2009). "Jean-Martin Charcot and movement disorders: neurological legacies to the 21st century". International Parkinson and Movement Disorder Society. Retrieved 2013.

Gusnard D., Akbudak E., Shulman G., Raichle M. (2001). Proceedings of the National Academy of Science, March 27, 2001 vol. 98 no. 7

Greyson, Bruce. 2006. "Near-Death Experiences and Spirituality." Zygon: Journal of Religion and Science.

Geschwind N. "Behavioural changes in temporal lobe epilepsy". Psychol Med. 1979.

Hatfield E. Love, Sex and Intimacy. New York: Harper Collins 1993.

Harribance CC. Sean Harribance: A psychic predicts the future. Port of Spain: Sean Harribance Institute; 1994.

Heaton JP, Adams MA. Update on central function relevant to sex: remodeling the basis of drug treatments for sex and the brain. Int J Impot Res 2003.

Hartmann, E. (1995). "Making connections in a safe place: Is dreaming psychotherapy?" Dreaming 5

Hartmann, E. (2011). The Nature and Functions of Dreaming, Oxford University Press; also Hartmann, E. (2012). "On the nature and functions of dreaming"; http://www.ceoniric.cl/english/articles/on_the_nature_and_functions.htm

Hayden, B., Pearson, J., & Platt, M. (2009). "Fictive reward signals in the anterior cingulate cortex". Science, 324, 948–950. doi:10.1126/science.1168488

Hobson, J. A., Pace-Schott, E. F., Stickbold, R. (2003). "Dreaming and the brain: toward a cognitive neuroscience of conscious states". In E.F. Pace-Schott, M. Solms, M. Blagrove, S. Harnad (Eds.), Sleep and Dreaming. New York, USA, Cambridge University Press

Hobson, J.A.; McCarley, R. 1977. "The brain as a dream state generator: an activation-synthesis hypothesis of the dream process". American Journal of Psychiatry, 134

Hobson, J. A. (2009). Journal Nature Reviews Neuroscience, Oct 2009

Hobson, J.A. (2009). "REM sleep and dreaming: towards a theory of protoconsciousness". Nature Reviews 10 (11): 803–813. doi:10.1038/nrn2716. PMID 19794431

Horton, C., Christopher J. A., et.al. (2009). Consciousness and Cognition 18 (3)

Hoss, R. (2005). Dream Language: Self-Understanding through Imagery and Color. Ashland: Innersource

Halligan PW. 2002. Phantom limbs: the body in mind. Cogn Neuropsychiatr.

Halligan PW, Fink GR, Marshal JC, Vallar G. 2003. Spatial cognition: evidence from visual neglect. Trends Cogn Sci.

Harribance" International Journal of Yoga Vol. 5, 2012.

Harrington A. Unfinished business: models of laterality in the nineteenth century. In: Davidson RJ, Hugdhal K, editors. Brain asymmetry. MIT Press; 1995.

Hoss Robert J. "The Neuropsychology of Dreaming: Studies and Observations" 2013.

Hécaen H, Ajuriaguerra J. 1952. L'Héautoscopie. In: Méconnassiances et hallucinations corporelles. Paris: Masson.

Hécaen H, Green A. 1957. Sur l'héautoscopie. Encephale.

Irwin HJ. 1985. Flight of mind: a psychological study of the out-of- body experience. Metuchen (NJ): Scarecrow Press.

Jung (1945). "On the Nature of Dreams" (1945). In Collected Works 8: The Structure and Dynamics of the Psyche.

Jung, C. (1971), The Portable Jung, J. Campbell (Ed.), New York: Viking Press.

Jung, C. G. (1973). Man and His Symbols, New York, Dell Publishing.

Krippner S, Persinger M. Evidence for enhanced congruence between dreams and distant target material during periods of decreased geomagnetic activity. J Sci Explor 1996

Komisaruk BR, Whipple B. Love as sensory stimulation: physiological consequences of its deprivation and expression. Psychoneuroendocrinology 1998

Kölmel HW. 1985. Complex visual hallucinations in the hemianopic field. J Neurol Neurosurg Psychiatry.

Lackner JR. 1988. Some proprioceptive influences on the perceptual representation of body shape and orientation. Brain.

Litovitz TA, Penafiel M,Krause D,Zhang D,Mullins JM.The roleoftemporal sensing in bioelectromagnetic effects. Bioelectromagnetics 1997.

Lagace N, St-Pierre LS, Persinger MA. Attenuation of epilepsy-induced brain dam- age in the temporal cortices of rats by exposure to LTP-patterned magnetic fields. Neurosci Lett 2009.

Lacoboni, M. and Dapretto, M. "The mirror neuron system and the consequences of its dysfunction" Nature 2005.

Lackner JR. 1992. Sense of body position in parabolic flight. Ann N Y Acad Sci.

Leube DT, Knoblich G, Erb M, Grodd W, Bartels M, Kircher TT. 2003. The neural correlates of perceiving one's own movements. Neuroimage.

Laidler Keith J. 2002, "Energy and The Unexpected", Oxford University Press

Lippman CW. 1953. Hallucination of physical duality in migraine. J Nerv Ment Dis.

Lobel E, Kleine J, Leroy-Wilig A. 1999. Functional MRI of galvanic vestibular stimulation. J Neurophysiol.

Lunn V. 1970. Autoscopic phenomena. Acta Psychiatr Scand 46(Suppl 219).

Mulligan BP, Hunter MD, Persinger MA. Effects of geomagnetic activity and atmospheric power variations on quantitative measures of brain activity: replication of the Azerbaijani studies. Adv Space Res 2010.

Moody, Paul. 1975. Life after Life: The Investigation of a Phenomenon—Survival of Bodily Death . Atlanta: Mockingbird Books.

Martens, P.R. 1994. "Near-Death Experiences in Out-of-Hospital Cardiac Arrest Survivors. Meaningful Pheneomena or just Fantasy of Death?" Resuscitation.

Melzack R. 1990. Phantom limbs and the concept of a neuromatrix. Trends Neurosci.

Metzinger T. 2003. Being no one. Cambridge (MA): MIT Press.

Mittelstaedt H, Glasauer S. 1993. Illusions of verticality in weightlessness. Clin Invest.

Neisser U. 1988. The five kinds of self-knowledge. Phil Psychol.

Newberg, A. "Cerebral blood flow changes associated with different meditation practices and perceived depth of meditation" Psychiatry Research: Neuroimaging 2010.

Onions CT. The Oxford Dictionary of English Etymology. New York: Oxford University Press 1966.

Persinger, "'I would kill in God's name' role of sex, weekly church attendance, report of a religious experience and limbic lability" Perceptual and Motor Skills 1997.

Persinger "Experimental simulation of the God experience" Neurotheology 2003.

Persinger, Corradini, Clement, Keaney, et al "Neurotheology and its convergence with neuroquantology" NeuroQuantology 2010.

Persinger, Koren and St-Pierre "The electromagnetic induction of mystical and altered states within the laboratory" Journal of Consciousness Exploration and Research 2010.

Persinger "Case report: A prototypical spontaneous 'sensed presence' of a sentient being and concomitant electroencephalographic activity in the clinical laboratory" Neurocase 2008.

Persinger and Saroka "Potential production of Hughlings Jackson's "parasitic consciousness" by physiologically-patterned weak transcerebral magnetic fields: QEEG and source localization" Epilepsy & Behavior 28 (2013).

Persinger. "The neuropsychiatry of paranormal experiences". J Neuropsychiatry Clin Neurosci 2001.

Persinger, M. "Billions of Human Brains Immersed Within a Shared Geomagnetic Field: Quantitative Solutions and Implications for Future Adaptations" The Open Biology Journal, 2013.

Persinger MA, Lavallee CF. Theoretical and experimental evidence of macroscopic entanglement between human brain activity and photon emissions; implications for quantum consciousness and future applications. J Cons Explor Res 2010

Persinger MA, Lavallee CF. The sum of N=N and the quantitative support for the cerebral holographic and electromagnetic configuration of consciousness. J Cons Stud 2012

Persinger MA. 10-20 Joules as a neuromolecular quantum in medicinal chemistry: an alternative to the myriad molecular pathways? Curr Med Chem 2010

Persinger MA. On the possible representation of the electromagnetic equivalents of all human memory within the earth's magnetic field: implications for theoretical biology. Theor Biol Insights 2008

Persinger, MA. "Schumann Resonance Frequencies Found Within Quantitative Electroencephalographic Activity: Implications for Earth-Brain Interactions" International Letters of Chemistry, Physics and Astronomy 2014.

Persinger, MA. "Quantitative Evidence for Direct Effects Between Earth-Ionosphere Schumann Resonances and Human Cerebral Cortical Activity" International Letters of Chemistry, Physics and Astronomy 2014.

Persinger, MA. "Terrestrial and lunar gravitational forces upon the mass of a cell: relevance to cell function" International Letters of Chemistry, Physics and Astronomy 2014.

Persinger, MA. "Dream ESP Experiments and Geomagnetic Activity" Journal of American Society for Psychical Research Vol 83, 1989.

Persinger, MA. and Saroka, KS. "Protracted parahippocampal activity associated with Sean Persinger MA, Krippner S. Dream ESP experiments and geomagnetic activity. J Am Soc Psychical Res 1989

Persinger "Experimental Facilitation of the Sensed Presence: Possible Intercalation between the Hemispheres Induced by Complex Magnetic Fields" Journal of Nervous and Mental Disease 2002.

Palmer J. 1978. The out-of-body experience: a psychological theory. Parapsychol Rev.

Persinger MA. Geophysical variables and behaviour: LXXI. Differential contribution of geomagnetic activity to paranormal experiences concerning death and crisis: An alternative to the ESP hypothesis. Percept Motor Skills 1993

Persinger MA, Saroka KS, Lavallee CF, Booth JN, Hunter MD, Mulligan BP, et al. Correlated cerebral events between physically and sensory isolated pairs of subjects exposed to yoked circumcerebral magnetic fields. Neuroscience Lett 2010

Persinger MA, Roll WG, Tiller SG, Koren SA, Cook CM. Remote viewing with the artist Ingo Swann: Neuropsychological profile, electroencephalographic correlates, magnetic resonance imaging (MRI) and possible mechanisms. Percept Motor Skills 2002

Plato, Phaedrus 370 BC

Revonsuo, A. (2000). "The reinterpretation of dreams: an evolutionary hypothesis of the function of dreaming". Behavioral Brain Science 23.

Ryan A. New insights into the links between ESP and geomagnetic activity. J Sci Explor, 2008

Roth, G and Dicke, U,(2005) Evolution of the Brain and the Intelligence, TRENDS in Cognitive Sciences.

Ruby P, Decety J. 2001. Effect of subjective perspective taking during simulation of action: a PET investigation of agency. Nat Neurosci.

Rizzolatti, G. & Fadiga, L. (1998) Grasping objects and grasping action meanings: The dual role of monkey rostroventral premotor cortex (area F5). Novartis Foundation Symposium 218

Rizzolatti, G., Fadiga, L., Gallese, V. & Fogassi, L. (1996) Premotor cortex and the recognition of motor actions. Social Cognitive and Affective Neuroscience 3

Rizzolatti, G., Fogassi, L. & Gallese, V. (2001) Neurophysiological mechanisms underlying the understanding and imitation of action. Nature Reviews Neuroscience 2

Rizzolatti G., Fogassi L. & Gallese V. (2004) Cortical mechanism subserving object grasping, action understanding and imitation. In: The cognitive neurosciences, 3rd edition, ed. M. S. Gazzaniga, A Bradford Book/MIT Press

Rizzolatti, G. & Arbib, M. A. (1998) Language within our grasp. Trends in Neurosciences 21 [aRC, LLH]

Rizzolatti, G., Camarda, R., Fogassi, L., Gentilucci, M., Luppino, G. & Matelli, M. (1988) Functional organization of inferior area 6 in the macaque monkey. II. Area F5 and the control of distal movements. Experimental Brain Research 71

Rizzolatti, G. & Luppino, G. (2001) The cortical motor system. Neuron 31 [LF] Rizzolatti, G. & Matelli, M. (2003) Two different streams form the dorsal visual system: Anatomy and functions. Experimental Brain Research 153

Rizzolatti, G. & Sinigaglia, C. (2008) Mirrors in the brain. How our minds share actions and emotions. Oxford University Press.

Rizzolatti, G. & Sinigaglia, C. (2010) The functional role of the parieto-frontal mirror circuit: Interpretations and misinterpretations. Nature Reviews Neuroscience 11

Ratnasuriya, R.H. "Joan of Arc, creative psychopath: is there another explanation?" Journal of The Royal Society of Medicine 1986.

Ramachandran, V. S. (2000) Mirror neurons and imitation learning as the driving force behind "the great leap forward" in human evolution. Edge69. [Available Online at: http://www.edge.org/3rd_culture/ramachandran/ramachandran_in- dex.html]

Ramachandran,V. S. (2009) The neurons that shaped civilization. Available at: http://www.ted.com/talks/vs_ramachandran_the_neurons_that_shaped_civilization.Html

Rocca, M. A., Tortorella, P., Ceccarelli, A., Falini, A., Tango, D., Scotti, G., Comi, G. & Fillipi, M. (2008) The "mirror-neuron system" in MS: A 3 tesla fMRI study. Neurology 70

Rochat, M. J., Caruana, F., Jezzini, A., Escola, L., Intskirveli, I., Grammont, F., Gallese, V., Rizzolatti, G. & Umiltà, M. A. (2010) Responses of mirror neurons in area F5 to hand and tool grasping observation. Experimental Brain Research 204

Rochat, M. J., Serra, E., Fadiga, L. & Gallese, V. (2008) The evolution of social cognition: Goal familiarity shapes monkeys' action understanding. Current Biology 18

Rochat, P. (1998) Self-perception and action in infancy. Experimental Brain Research 123

Rosenbaum, D. (1991) Human motor control. Academic Press.

Roth, T. L. (2012) Epigenetics of neurobiology and behavior during development and adulthood. Developmental Psychobiology 54. doi: 10.1002/dev.20550.

Rushworth, M. F., Mars, R. B. & Sallet, J. (2013) Are there specialized circuits for social cognition and are they unique to humans? Current Opinion in Neurobiology 23

Russell, J. L., Lyn, H., Schaeffer, J. A. & Hopkins, W. D. (2011) The role of socio- communicative rearing environments in the development of social and physical cognition in apes. Developmental Science 14

Sampson, G. (2002) Exploring the richness of the stimulus. The Linguistic Review 19

Sanefuji, W. & Ohgami, H. (2013) "Being-imitated" strategy at home-based intervention for young children with autism. Infant Mental Health Journal 34. doi: 10.1002/imhj.21375.

Santos, L. R., Nissen, A. G. & Ferrugia, J. A. (2006) Rhesus monkeys, Macaca mulatta, know what others can and cannot hear. Animal Behaviour 71

Singer, T. et. al. "Empathy for Pain Involves the Affective but not Sensory Components of Pain" Science 303 (2004).

Sheils D. 1978. A cross-cultural study of beliefs in out-of-the-body experiences, waking and sleeping. J Soc Psych Res.

Strassman, R. "DMT: The Spirit Molecule" 2001.

Smith BH. 1960. Vestibular disturbances in epilepsy. Neurol.

Schrodinger Erwin. (2012) "What is Life?: With Mind and Matter and Autobiographical Sketches", Cambridge University Press

Schumann W. O. (1952). "Über die strahlungslosen Eigenschwingungen einer leitenden Kugel, die von einer Luftschicht und einer Ionosphärenhülle umgeben ist". Zeitschrift und Naturfirschung 7a. Bibcode:1952ZNatA...7..149S. doi:10.1515/zna-1952-0202.

Schumann W. O. (1952). "Über die Dämpfung der elektromagnetischen Eigenschwingnugen des Systems Erde – Luft – Ionosphäre". Zeitschrift und Naturfirschung 7a. Bibcode:1952ZNatA...7..250S. doi:10.1515/zna-1952-3-404.

Schumann W. O. (1952). "Über die Ausbreitung sehr Langer elektriseher Wellen um die Signale des Blitzes". Nuovo Cimento 9. doi:10.1007/BF02782924.

Schumann W. O. & H. König (1954). "Über die Beobactung von Atmospherics bei geringsten Frequenzen". Naturwiss 41. Bibcode:1954NW.41.183S. doi:10.1007/BF00638174.

Shipley JT. Dictionary of Word Origins. New York: Philosophical Library 1945.

Stefano GB, Scharrer B. Endogenous morphine and related opiates, a new class of chemical messengers. Adv Neuroimmunol 1994

Stefano GB, Scharrer B, Smith EM, Hughes TK, et al. Opioid and opiate immunoregulatory processes. Crit Rev in Immunol 1996.

Simpson JA, Rholes WS. Stress and secure base relationships in adulthood. In: Bartholomew K, Perlman D, editors. Advances in personal relationships (Vol. 5): Attachment processes in adulthood. London: Kingsley 1994

Slingsby BT, Stefano GB. Placebo: Harnessing the power within. Modern Aspects of Immunobiology 2000

Slingsby BT, Stefano GB. The active ingredients in the sugar pill: Trust and belief. Placebo 2001

Small DM, Jones-Gotman M, Dagher A. Feeding-induced dopamine release in dorsal striatum correlates with meal pleasantness ratings in healthy human volunteers. Neuroimage 2003

Small DM, Zatorre RJ, Dagher A, Evans AC, Jones-Gotman M. Changes in brain activity related to eating chocolate: From pleasure to aversion. Brain 2001

Smith CM. Elements of Molecular Neurobiology. 3rd ed. New York: Wiley-Liss 2002

Sonetti D, Peruzzi E, Stefano GB. Endogenous morphine and ACTH association in neural tissues. Medical Science Monitor 2005

Spector S, Munjal I, Schmidt DE. Endogenous morphine and codeine. Possible role as endogenous anticonvulsants. Brain Res 2001

Spencer H. Principles of Psychology. New York: Appleton 1800

Stefano GB. Endocannabinoid immune and vascular signaling. Acta Pharmacologica Sinica 2000

Stefano GB, Benson H, Fricchione GL, Esch T. The Stress Re- sponse: Always good and when it is bad. New York: Medical Science International 2005

Stefano GB, Cadet P, Zhu W, Rialas CM, Mantione K, Benz D et al. The blueprint for stress can be found in invertebrates. Neuroendocrinology Letters 2002.

Sternberg, Robert J. (2007). "Triangulating Love". In Oord, T. J. The Altruism Reader: Selections from Writings on Love, Religion, and Science. West Conshohocken, PA: Templeton Foundation. ISBN 9781599471273.

Sternberg, Robert J. (2004). "A Triangular Theory of Love". In Reis, H. T.; Rusbult, C. E. Close Relationships. New York: Psychology Press. ISBN 0863775950.

Sternberg, Robert J. (1997). "Construct validation of a triangular love scale". European Journal of Social Psychology 27

Slater E, Beard AW. The schizophrenia-like psychoses of epilepsy. Br J Psychiatry 1963.

Tolstoy Leo, Anna Karenina 1877

Tan SY, Shigaki D (May 2007). "Jean-Martin Charcot (1825–1893): pathologist who shaped modern neurology". Singapore Med J 48. PMID 17453093

Tommasi MCO. Orgiasmo orgies and ritual in the ancient world: a few notes. Kervan 2006-2007

Turner, J. H. (2000b). On the origins of human emotions: A sociological inquiry into the evolution of human affect. Stanford, California: Stanford University Press.

Venkatasubramanian G, Jayakumar PN, Nagendra HR, Nagaraja D, Deeptha R, Gangadhar BN. Investigating paranormal phenomena: Functional brain imaging of telepathy. Int J Yoga 2008

Whinnery, J.E. 1997. "Psychophysiologic Correlates of Unconsciousness and near- death experiences." Journal of Near-Death Studies.

Watkins, K. & Paus, T. Modulation of motor excitability during speech perception: the role of Broca's area. J. Cogn. Neurosci.

Zacks JM, Ollinger JM, Sheridan MA, Tversky B. 2002. A parametric study of mental spatial transformations of bodies. Neuroimage.

The Art of Neuroscience in Everything

Printed in Dunstable, United Kingdom